# EXPERIENCE IN THE DEVELOPMENT AND APPLICATION OF LEVEL 2 PROBABILISTIC SAFETY ASSESSMENT FOR NUCLEAR POWER PLANTS

The following States are Members of the International Atomic Energy Agency:

AFGHANISTAN
ALBANIA
ALGERIA
ANGOLA
ANTIGUA AND BARBUDA
ARGENTINA
ARMENIA
AUSTRALIA
AUSTRIA
AZERBAIJAN
BAHAMAS
BAHRAIN
BANGLADESH
BARBADOS
BELARUS
BELGIUM
BELIZE
BENIN
BOLIVIA, PLURINATIONAL STATE OF
BOSNIA AND HERZEGOVINA
BOTSWANA
BRAZIL
BRUNEI DARUSSALAM
BULGARIA
BURKINA FASO
BURUNDI
CABO VERDE
CAMBODIA
CAMEROON
CANADA
CENTRAL AFRICAN REPUBLIC
CHAD
CHILE
CHINA
COLOMBIA
COMOROS
CONGO
COOK ISLANDS
COSTA RICA
CÔTE D'IVOIRE
CROATIA
CUBA
CYPRUS
CZECH REPUBLIC
DEMOCRATIC REPUBLIC OF THE CONGO
DENMARK
DJIBOUTI
DOMINICA
DOMINICAN REPUBLIC
ECUADOR
EGYPT
EL SALVADOR
ERITREA
ESTONIA
ESWATINI
ETHIOPIA
FIJI
FINLAND
FRANCE
GABON
GAMBIA, THE
GEORGIA
GERMANY
GHANA
GREECE
GRENADA
GUATEMALA
GUINEA
GUYANA
HAITI
HOLY SEE
HONDURAS
HUNGARY
ICELAND
INDIA
INDONESIA
IRAN, ISLAMIC REPUBLIC OF
IRAQ
IRELAND
ISRAEL
ITALY
JAMAICA
JAPAN
JORDAN
KAZAKHSTAN
KENYA
KOREA, REPUBLIC OF
KUWAIT
KYRGYZSTAN
LAO PEOPLE'S DEMOCRATIC REPUBLIC
LATVIA
LEBANON
LESOTHO
LIBERIA
LIBYA
LIECHTENSTEIN
LITHUANIA
LUXEMBOURG
MADAGASCAR
MALAWI
MALAYSIA
MALI
MALTA
MARSHALL ISLANDS
MAURITANIA
MAURITIUS
MEXICO
MONACO
MONGOLIA
MONTENEGRO
MOROCCO
MOZAMBIQUE
MYANMAR
NAMIBIA
NEPAL
NETHERLANDS, KINGDOM OF THE
NEW ZEALAND
NICARAGUA
NIGER
NIGERIA
NORTH MACEDONIA
NORWAY
OMAN
PAKISTAN
PALAU
PANAMA
PAPUA NEW GUINEA
PARAGUAY
PERU
PHILIPPINES
POLAND
PORTUGAL
QATAR
REPUBLIC OF MOLDOVA
ROMANIA
RUSSIAN FEDERATION
RWANDA
SAINT KITTS AND NEVIS
SAINT LUCIA
SAINT VINCENT AND THE GRENADINES
SAMOA
SAN MARINO
SAUDI ARABIA
SENEGAL
SERBIA
SEYCHELLES
SIERRA LEONE
SINGAPORE
SLOVAKIA
SLOVENIA
SOMALIA
SOUTH AFRICA
SPAIN
SRI LANKA
SUDAN
SWEDEN
SWITZERLAND
SYRIAN ARAB REPUBLIC
TAJIKISTAN
THAILAND
TOGO
TONGA
TRINIDAD AND TOBAGO
TUNISIA
TÜRKİYE
TURKMENISTAN
UGANDA
UKRAINE
UNITED ARAB EMIRATES
UNITED KINGDOM OF GREAT BRITAIN AND NORTHERN IRELAND
UNITED REPUBLIC OF TANZANIA
UNITED STATES OF AMERICA
URUGUAY
UZBEKISTAN
VANUATU
VENEZUELA, BOLIVARIAN REPUBLIC OF
VIET NAM
YEMEN
ZAMBIA
ZIMBABWE

The Agency's Statute was approved on 23 October 1956 by the Conference on the Statute of the IAEA held at United Nations Headquarters, New York; it entered into force on 29 July 1957. The Headquarters of the Agency are situated in Vienna. Its principal objective is "to accelerate and enlarge the contribution of atomic energy to peace, health and prosperity throughout the world".

IAEA-TECDOC-2081

# EXPERIENCE IN THE DEVELOPMENT AND APPLICATION OF LEVEL 2 PROBABILISTIC SAFETY ASSESSMENT FOR NUCLEAR POWER PLANTS

## SUMMARY OF A TECHNICAL MEETING

INTERNATIONAL ATOMIC ENERGY AGENCY
VIENNA, 2025

# COPYRIGHT NOTICE



For further information on this publication, please contact:

Safety Assessment Section
International Atomic Energy Agency
Vienna International Centre
PO Box 100
1400 Vienna, Austria
Email: Official.Mail@iaea.org


Printed by the IAEA in Austria
February 2025
https://doi.org/10.61092/iaea.f6ss-47dp

**IAEA Library Cataloguing in Publication Data**

Names: International Atomic Energy Agency.
Title: Experience in the development and application of level 2 probabilistic safety assessment for nuclear power plants / International Atomic Energy Agency.
Description: Vienna : International Atomic Energy Agency, 2025. | Series: IAEA TECDOC series, ISSN 1011-4289 ; no. 2081 | Includes bibliographical references.
Identifiers: IAEAL 25-01738 | ISBN 978-92-0-104625-3 (paperback : alk. paper) | ISBN 978-92-0-104525-6 (pdf)
Subjects: LCSH: Nuclear power plants — Safety measures. | Nuclear power plants — Risk assessment. | Nuclear facilities — Safety regulations.

# FOREWORD

One of the major concerns since the beginning of the nuclear power industry has been the potential impact of accidental radioactive releases on the population and the environment. The estimation of the likelihood of occurrence of such accidental radioactive releases and the calculation of the consequences they could generate have been the focus of the probabilistic approach since it was first developed for nuclear power plants in the 1950s. Following the initial studies, probabilistic safety assessment (PSA) was divided into three levels, with Level 2 PSA focusing on determining the frequency of radioactive releases outside the reactor containment and the characteristics of the source term. Despite the complexity and the uncertainty of the phenomena, substantial progress has been achieved in Member States in recent years regarding the development and use of Level 2 PSA for nuclear power plants. That progress is supported by the results of research programmes on severe accident phenomena and on the behaviour of structures, systems and components facing the conditions created by severe accidents.

This publication provides a summary of the presentations and the results of discussions at the Technical Meeting on Experience in the Development and Application of Level 2 Probabilistic Safety Assessment for Nuclear Power Plants held from 4 to 7 May 2021. It presents information on the latest practices and experiences in Member States with regard to the development and application of Level 2 PSA for nuclear power plants. It also summarizes suggestions provided during the technical meeting, which were used, as appropriate, in the revision process for IAEA Safety Standards Series No. SSG-4, Development and Application of Level 2 Probabilistic Safety Assessment for Nuclear Power Plants.

The IAEA wishes to acknowledge the support provided by the contributors and reviewers listed at the end of this publication, in particular those who provided presentations. The IAEA officers responsible for this publication were J. Luis Hernandez, R. Minibaev and Z. Stone of the Division of Nuclear Installation Safety.

*EDITORIAL NOTE*

CONTENTS

# 1. INTRODUCTION

## 1.1. BACKGROUND

IAEA Safety Standards Series Nos SSG-3, Development and Application of Level 1 Probabilistic Safety Assessment for Nuclear Power Plants[1], and SSG-4, Development and Application of Level 2 Probabilistic Safety Assessment for Nuclear Power Plants [2] were developed to provide recommendations for the development and application of Level 1 and Level 2 probabilistic safety assessment (PSA) for nuclear power plants (NPPs) respectively and to meet the relevant requirements established in the IAEA safety standards existing at that time, in particular:

- GSR Part 4, Safety Assessment for Facilities and Activities;
- SSR-2/1, Safety of Nuclear Power Plants: Design;
- SSR-2/2, Safety of Nuclear Power Plants: Operation.

All these Safety Requirements publications have been revised, taking into account the latest developments and relevant practices in the Member States as well as the feedback from the accident at the Fukushima Daiichi NPP in 2011. SSG-3 was recently revised [1], and SSG-4 [2] is currently under revision, and will be published as SSG-4 (Rev. 1) [3].

Among the significant changes incorporated into the requirements established in IAEA Safety Standards Series Nos GSR Part 4 (Rev. 1) [4], SSR-2/1 (Rev. 1) [5] and SSR-2/2 (Rev. 1) [6] are those related to severe accident management, including the use of non-permanent equipment, and the margins to both withstand levels of natural hazards more severe than those considered for design, derived from the hazard evaluation for the site, and to avoid cliff edge effects. Those changes have had an impact on the safety provisions incorporated in the plant design as well as on the plant operation for all plant states, to cope with severe accidents which are considered in Level 2 PSA.

The development and application of Level 2 PSA for NPPs have been substantially consolidated and progressed in Member States in recent years. In particular, the lessons learned from major accidents have been largely applied to update severe accident management guidelines and Level 2 PSA studies.

Moreover, the latest applications of Level 2 PSA to assess the overall level of safety of NPPs have considered the interactions and dependencies in a multi-unit context, highlighting the need to consider the potential impact of internal and external hazards and their combinations as well as the common cause failures and human factors among different units at the site (for specific recommendations see IAEA Safety Standards Series No. SSG-89, Evaluation of Seismic Safety for Nuclear Installations [7]; further information is provided in Ref. [8]) .

In addition, Level 2 PSA has been applied to evaluate the contribution and challenges of using non-permanent equipment to cope with severe accident situations to ensure a high level of safety of NPPs.

[1] INTERNATIONAL ATOMIC ENERGY AGENCY, Development and Application of Level 1 Probabilistic Safety Assessment for Nuclear Power Plants, IAEA Safety Standards Series No. SSG-3, IAEA, Vienna (2010): now superseded by SSG-3 (Rev. 1) [1].

In recent years, extensive research programmes have been conducted to improve knowledge and reduce uncertainties related to the phenomena associated with severe accidents. The results have been used to substantiate the implementation of specific design solutions and strategies for both new and existing NPPs to cope with severe accidents and to improve and validate code simulation capabilities. Therefore, Level 2 PSA models have been applied to provide insights on possible prioritization for those research programmes.

Given the interest of Member States in this area, IAEA organized a virtual technical meeting on Experience in the Development and Application of Level 2 Probabilistic Safety Assessment for Nuclear Power Plants (herein referred to as 'the technical meeting'), from 4 to 7 May 2021. The purpose of the technical meeting was to provide a forum for discussion on current experience of Member States with a focus on challenges and issues, such as: the definition of risk metrics for Level 2 PSA, modelling of non-permanent equipment in a Level 2 PSA context, and the assessment of multi-unit and multi-source considerations for Level 2 PSA.

## 1.2. OBJECTIVES

The main purpose of this TECDOC is to capture the current state of knowledge and practices gained in the last years in Member States concerning the development and use of Level 2 PSA, in order to disseminate this knowledge and these practices to a larger audience.

Additionally, this TECDOC summarizes the key insights from the discussions held during the technical meeting, and the path forward in the Level 2 PSA area.

Finally, this TECDOC captures the technical meeting participants' suggestions for the revision of SSG-4 [2], adding transparency to the process of reviewing the Safety Guide.

## 1.3. SCOPE

This TECDOC summarizes the technical sessions on regulatory, research and development, and industry perspectives during the technical meeting related to the development and the application of Level 2 PSA for NPPs, as follows:

- Assessment of the safety level improvements related to:
  - Incorporation in the NPP design of additional safety features considered for design extension conditions;
  - Use of non-permanent equipment;
  - Multiunit considerations.
- Assessment of advantages and disadvantages of strategies for dealing with core melt and with damage of fuel stored in the spent fuel pool inside the reactor containment, and their related phenomena.
- Assumptions and factors related to modelling of human performance in Level 2 PSA.
- Prioritization of research conducted in support of severe accident strategies and improvements of severe accident simulation codes capabilities.
- Enhancement and update of severe accident management guidelines using insights from Level 2 PSA.
- Current practices considering low power and shutdown states as well as internal and external hazards, and their combinations in Level 2 PSA.
- Development and application of Level 2 PSA for NPPs with advanced reactor technologies (e.g. small modular reactors (SMRs), non-light-water reactors (non-LWR)).

This publication also includes summaries of discussions held during the technical meeting which enabled participants to make suggestions for IAEA future activities.

## 1.4. STRUCTURE

Section 1 presents the background, scope and structure of this TECDOC. Section 2 provides a summary of the technical and discussion sessions during the technical meeting. Each summary is based on the discussion summary presented and agreed upon by the meeting participants. Section 3 provides the general technical meeting conclusions agreed by the participants to the technical meeting. Section 4 provides a conclusion of the activities and discussion on the coordinated efforts regarding Level 2 PSA approaches.

# 2. SUMMARY OF THE TECHNICAL MEETING SESSIONS

## 2.1. SESSION I: OPENING SESSION

The opening session focused on providing an overview of current and future activities and publications under development at the IAEA in relation to PSA. The activities highlighted were the main lessons learned from recent Technical Safety Review services, in particular those in PSA, and the planned Technical Safety Review missions. This information allowed participants to get the benefits of such services for enhancing the PSA studies under review to comply with relevant IAEA Safety Requirements and recommendations.

Regarding the current activities and publications under development, the participants were updated on the status of relevant PSA documents, such as: the revision of the Safety Guides on PSA; current Safety Reports, such as the one on human reliability analysis (HRA) for nuclear installations and multi-unit PSA aspects; and other publications, such as TECDOCs on integrated risk informed decision making [9], seismic PSA [10], full scope Level 1 PSA for NPPs [11], and advanced PSA for NPPs (in preparation).

## 2.2. SESSION II: REGULATORY PERSPECTIVE

### 2.2.1. Session II description

The first technical session was dedicated to representing regulatory perspectives regarding current status of practices on Level 2 PSA. For regulatory bodies, the issues of a reasonable definition of risk metrics and safety goals are more acute than for PSA developers. The criteria adopted needs to be justified and intuitive. In this section, regulatory bodies from Pakistan, the Russian Federation, Türkiye, Canada, Finland and Argentina presented their experience and approaches in Level 2 PSA. The abstracts of presentations are presented in Sections 2.2.2–2.2.6. The discussion is presented in Section 2.2.7.

### 2.2.2. Development of the regulator's Level 2 probabilistic safety assessment model in Pakistan

The Pakistan Nuclear Regulatory Authority has developed an independent Level 1 PSA model for 300 MW(e) NPPs which is being utilized for various regulatory applications, e.g. review of preliminary safety analysis reports and final safety analysis reports, review of modifications in design and/or technical specifications, prioritization of regulatory inspections, risk informed decision making, etc. The Level 1 PSA model is used as the basis for the development of the

Level 2 PSA model. This presentation described the development process of the Level 2 PSA model, the challenges faced during its development, and potential future applications.

The development of Level 2 PSA was started with developing the interface between Level 1 PSA and Level 2 PSA models. For this purpose, the core damage sequences from Level 1 PSA were analysed in containment safety event trees (or bridge trees) to resolve the unknown status of safety systems and to include the containment safety systems required for severe accident mitigation. The bridge trees transformed hundreds of core damage sequences from Level 1 PSA to a manageable number of plant damage states (PDS) for further analysis in Level 2 PSA. After that, containment event trees (CETs) were developed for analysis of severe accident phenomena arising from PDS for identification of major containment failure modes along with their frequencies as well as estimation of magnitudes and frequencies of radioactive releases to the environment. The end states of CETs will provide the major contributors to early containment failure (including containment bypass events) and the frequency of different release categories. These release categories will be further analysed for calculation of associated source terms to determine the quantity of radioactive material released from the plant to the environment. The development of the regulator's independent Level 2 PSA model is a challenging task which requires expertise in plant systems, operational aspects, severe accident mitigation strategies, supporting deterministic analysis, etc., and a dedicated team with competence in diverse disciplines. The development of Level 2 PSA will enhance the capability of the Pakistan Nuclear Regulatory Authority PSA team regarding analysis of severe accident progression phenomena and relevant severe accident mitigation actions and will enable it to carry out independent assessment of licensee's submissions.

### 2.2.3. The use of Level 2 probabilistic safety assessment in nuclear power plant safety regulation in the Russian Federation

The system of state regulation in the field of atomic energy use in the Russian Federation provides the basis for the NPP operating organization to obtain licences for the following activities for each NPP unit: siting, construction, operation, and decommissioning. In addition, during the operation of each NPP unit, operating organizations plan to perform a periodic safety assessment of each NPP unit every ten years. In all these cases, with the exception for the siting and decommissioning phases, when applying to the state regulatory body (Rostechnadzor) for obtaining the appropriate licence, a report with the results of Level 1 and Level 2 PSA of each NPP unit is required.

Currently the Russian regulatory framework in the field of atomic energy use in terms of PSA consists of the federal rules and regulations (high level regulatory documents containing requirements), as well as safety guides for the use of atomic energy (lower level regulatory documents containing recommendations).

In accordance with the requirements of Refs [12, 13], NPP safety analysis reports have to provide a Level 2 PSA performed for all initiating events (IEs), all modes of normal operation, all locations of nuclear material, radioactive substances and radioactive waste at each NPP unit.

In accordance with the recommendations of Ref. [14], the objectives of Level 2 PSA are:

- Determination of total probability of a large accidental release for each NPP unit at intervals of one year for all IEs, all modes of normal operation, all locations of nuclear materials, radioactive substances and radioactive waste at each NPP unit;

- Determination of compliance/non-compliance of total probability of a large accidental release for each NPP unit with the target 1.0 x $10^{-7}$ per reactor-year established by the federal rules and regulations in the field of atomic energy use Ref. [12];
- Determination of the categories of accidental releases for each NPP unit and the consequences of accidents determined by the accidental releases of each such category;
- Identification of factors that have the highest impact on the consequences of accidents;
- Determination on this basis of additional measures to ensure the safety of each NPP unit.

To date, the first stage has been completed for the units of all Russian NPPs: Level 2 PSAs are submitted to Rostechnadzor as part of the sets of documents justifying the safety of NPP units, for internal IEs when operating at rated power for all NPP units in operation.

The presentation covered the topics related to the implementation of Level 2 PSA by operating organizations for old and new generation units, the peculiarities of using the results of deterministic calculations performed in support of Level 2 PSA, as well as issues related to the applicability of the principle of exclusion from consideration in the design of events with severe radiation consequences based on their physical impossibility or low probability with a high level of confidence (as discussed in SSR-2/1 (Rev. 1) [5]).

### 2.2.4. Assessment of early release risk in Finland

Assessment of the large release risk and early release risk is an important part of the substantiation of the practical elimination concept in the overall safety philosophy in Finland, and Level 2 PSA plays a central role in this context. Reference [15] states that:

> "The release of radioactive substances arising from a severe accident shall not necessitate large scale protective measures for the public nor any long-term restrictions on the use of extensive areas of land and water. In order to restrict long-term effects, the limit for the atmospheric release of Cesium-137 is 100 terabecquerel (TBq) [1E+12 Bq]. The possibility of exceeding the set limit shall be extremely small. The possibility of a release requiring measures to protect the public in the early stages of the accident shall be extremely small."

Reference [16] implements the above requirement in terms of requirements for Level 2 PSA and associated safety goals. Full-scope Level 1 and 2 PSAs are required both as part of the construction licence application and operating licence application for a new NPP. The guide also requires updating of PSA and various risk informed applications during operation of an NPP.

With regard to Level 2 PSA, the following numerical acceptance criteria, which are target values for operating plants, are set as shown in Figure 1:

(a) The large release frequency (LRF) has to be less than 5 x $10^{-7}$ per reactor-year. The definition for 'large release' is taken from the Government Decree, i.e. it means a release more than 1E+14 Bq Cesium-137.

(b) The accident sequences, in which the containment function fails or is lost in the early phase of a severe accident, have only a small contribution to the reactor core damage frequency (CDF). "Early" means that there is no time to implement the warning and protective measures prior to the release. An exact number of hours has not been defined but warning and protection are typically estimated to take approximately four hours after the rescue department receives information on the need to take shelter. The objective is that protective measures are not needed in a situation in which there would practically be no time to implement them.

The release assessments need to take into account all of the spent nuclear fuel located at the plant unit, which means, for example that the spent fuel pool located at the unit needs to also be included in the assessment.

The fulfilment of these risk criteria is reviewed as part of the licensing process for new NPPs in Finland: Olkiluoto 3 NPP, which is now in commissioning phase and Hanhikivi 1 NPP, which is now in the construction licence application phase. Concerning operating NPPs (Loviisa 1 and 2, and Olkiluoto 1 and 2), the fulfilment of the numerical targets has been reviewed as part of the periodic safety reviews.

Loviisa 1 and 2 and Olkiluoto 1 and 2, each of which has been in operation about 40 years, have implemented several significant safety improvements during past years. Currently, they approximately meet the targets values for Level 1 PSA (CDF < $1 \times 10^{-5}$ per reactor-year), but do not meet the target values for the large release frequency[2]. Loviisa 1 and 2 can be said to meet the early release frequency target. This result can be explained by the severe accident management strategy developed for the plant, especially technical solutions for securing in-vessel retention, and large primary circuit water inventory, meaning that severe accident sequences do not typically lead to an early release in Loviisa NPP. Regarding Olkiluoto 1 and 2, the fraction of early releases frequency is not small, which can be explained by the design with a smaller primary circuit and containment volumes and strong dependency on electricity driven safety functions. For both Loviisa 1 and 2 and Olkiluoto 1 and 2, decreasing CDF has been concluded to be the most effective way to further decrease large release and early release risk.

Olkiluoto 3, which is a generation III light water reactor, fulfils all numerical risk criteria. Nevertheless, according to the principle of continuous improvement of safety, means to practically eliminate early and large releases need to be considered even for Olkiluoto 3. The design solution for ensuring residual heat removal in case of loss of internal alternating current power supply is considered acceptable[3]. This plant modification will be implemented after the start of commercial operation, and it originates from the Forsmark incident in 2006.

PSA for Hanhikivi 1 NPP has not yet been under regulatory review.

---

[2] "Approximately" here is understood as those NPPs were built according to previous standards to meet previous risk metrics, therefore the updated risk metrics are consequently not totally attainable for those NPP designs.

[3] This refers to the use of passive safety features for ensuring the containment heat removal function during severe accidents.

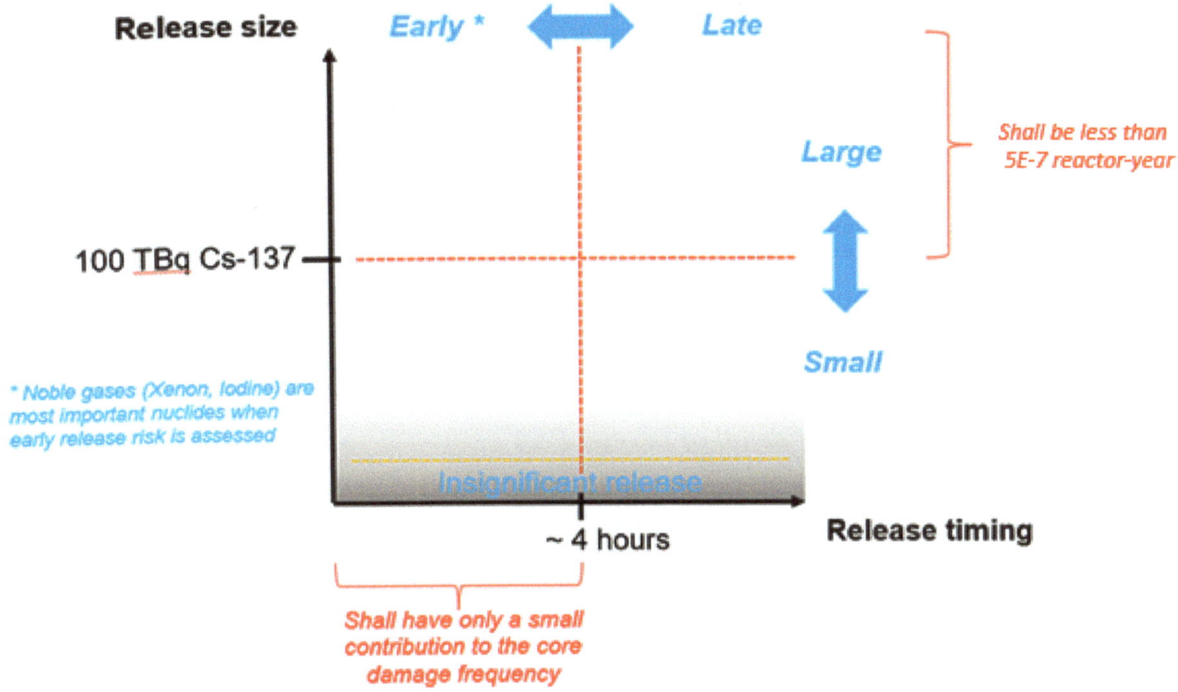


*FIG. 1. Proposal definition of 'large' or 'early' release in Finland (courtesy of J. Holmberg, Radiation and Nuclear Safety Authority, Finland).*

### 2.2.5. Probabilistic approach for nuclear safety regulation in Argentina

The Argentine Nuclear Regulatory Authority (ARN) has established clear criteria for acceptable potential exposure risks, utilizing a probabilistic approach to individual radiological risk. This approach aligns with the dose limitation system philosophy of the International Commission on Radiological Protection, which is used for radiation protection purposes.

The goal is to restrict individual risk from potential exposures to levels similar to those from normal operational exposures. The ARN evaluates the safety level of a facility using a regulatory tool known as the 'acceptability criterion curve'. More information can be found in Ref. [17].

### 2.2.6. Regulatory review of Level 2 probabilistic safety assessment for multi-unit CANDU plants in Canada

Reference [18] sets out the requirements with respect to the PSA. It requires that the PSA needs to include:

- Level 1 and Level 2 PSAs;
- Internal and external events and their potential combinations;
- Multi-unit impacts (if applicable);
- Radioactive sources other than the reactors;
- Other plant operating states (POS) not covered by at-power and shutdown states;
- Importance, uncertainty and sensitivity analyses.

Reference [18] requires the licensees to seek CNSC acceptance of the PSA methodology prior to the conduct of the PSA, and refers to SSG-3 and SSG-4 [2], and to Ref. [19], for guidance. The safety goals are defined in Ref. [20]. The Level 2 safety goals include:

> **“Small Release Frequency:** Sum of frequencies of all sequences leading to a release of more than 1E+15 Bq of I-131 shall be less than 1 x $10^{-5}$ per reactor-year. A higher release may trigger temporary evacuation of the local population.
>
> **“Large Release Frequency:** Sum of frequencies of all sequences leading to a release of more than 1E+14 Bq of Cs-137 shall be less than 1 x $10^{-6}$ per reactor-year. A higher release may trigger long-term relocation of the local population.”

For existing reactors, the numerical safety goals are defined by the licensees with a frequency one order of magnitude higher, in accordance with Ref. [21], and by setting safety goal targets in accordance with Ref. [20].

*2.2.6.1. Scope and level of detail of Level 2 PSA*

The first step in the development of a Level 2 PSA consists of revisiting the initiating events screening in the Level 1 PSA to check if there are no IEs that were screened out from the scope of the Level 1 PSA, which could result in a large radioactive release.

As per CANDU practice, PSAs are developed on a per-unit basis; however, multi-unit impacts are duly accounted for. The scope of the Level 2 PSA includes all IEs, all plant operating states, and all radioactive sources. However, alternate assessment methods are allowed for external hazards, and radioactive sources other than the reactor core. PSA models are developed consistent with the risk impact. Hazards screening and hazards combinations analyses are submitted in separate reports, or as part of the seismic PSA for the seismic induced internal fires and internal floods. A screening analysis is conducted for various non-reactor sources of radioactivity. As an example, the analysis of spent fuel pool shows that the time for the pool water to boil, or for the top row of bundles to become uncovered, is too long, and therefore these can be screened out as their contribution to LRF is negligible. The scope of the Level 2 PSA, as per the current practice, include:

- Detailed Level 2 PSA for internal events at power state;
- Simplified Level 2 PSAs for internal events at shutdown state;
- Simplified Level 2 PSA at power for internal fires, internal floods, seismic events, and high winds;
- Level 2 at-shutdown PSAs for internal fires, internal flood, seismic events, and high winds are excluded from the scope as their contribution to the large release frequency is judged to be negligible.

The site LRF for multi-unit CANDU plants is calculated using a careful aggregation of the per-unit LRF results. The minimal cut sets for the single (reference) unit are integrated to identify single, and multi-unit sequences. This step is important prior to the risk aggregation to avoid double (multiple) counting of accidents sequences. A multi-unit bridging tree is also developed with simplified replicated models for non-reference units.

Aggregation across hazards is performed using simple summation, and some simplified approaches are used to estimate the LRF for the cases where a Level 2 PSA is not fully developed (internal floods, internal fires, seismic, and high wind PSA).

#### 2.2.6.2. *Regulatory review of Level 2 PSA*

As a pre-requisite for the Level 2 PSA review, the PSA methodology and computer codes need to be accepted prior to the conduct of the PSA. Similarly, Level 1 PSA also needs to be reviewed and judged to be compliant, given this is the building block for Level 2 PSA.

The review criteria used for the acceptance of the Level 2 PSA methodology, are based on the SSG-4 [2], and CSA N290-17 [19]. Additional international references are used for the review of the detailed Level 2 PSA reports. These include Refs [22, 23], NUREGs and EPRI reports as referenced in the accepted PSA methodology.

The regulatory review process is conducted in two stages:

- **Stage 1 review (high level review):** This is a qualitative review aiming to assess the overall PSA documentation, consistency with the accepted methodology, to demonstrate that safety goals are met, and to identify key accident sequences and areas for the Stage 2 review;

- **Stage 2 review (technical review):** This is a quantitative and detailed review based on spot check of a few samples of the Level 2 PSA tasks (Level 1/Level 2 interface, containment fault trees, accident progression event trees (APET), severe accident analyses, assignment of release category, human interactions, Level 2 PSA model integration, sensitivity, uncertainty and importance analyses).

Regular exchange meetings with licensees also support the regulatory review.

The focus, in stage 2 of the regulatory review, regarding the PSA task 'Level 1/Level 2 interface', is on the attributes used for the grouping the Level 1 end states into PDSs, and specifically the PDS for multi-unit sequences in multi-unit CANDU plants. PDS attributes include the consideration of the type of IE, severity and timing of the accident sequence, separation of sequences leading to releases inside and outside the containment, separation of single-unit sequences and multi-unit accident sequences, and the consideration of pre-existing and on-demand containment failures using the bridging event tree. The emphasis of the review is on the methodology used for the selection of representative accident sequences for each PDS considering both frequency and impact of accident sequences. It also includes the review of the accident simulation with the MAAP-CANDU code [24].

In the APET task, the focus is on the development of nodal questions with special emphasis on the in-vessel retention (IVR) strategy, the understanding of severe accident phenomenology and its impact on the containment, as well as on the assignment of APET end states to different release categories.

#### 2.2.6.3. *Level 2 PSA regulatory applications and review challenges*

The results of the review of the overall Level 2 PSA model integration are used to:

- Demonstrate that Level 2 safety goals are met;
- Identify major containment failure modes and their frequencies;
- Gain insights into the progression of severe accidents, and identify plant-specific challenges and vulnerabilities to severe accidents and support development of SAMGs;
- Provide input for emergency planning;
- Support the Emergency Operating Centre for prognostic analyses.

The challenges associated with the regulatory review of CANDU Multi-unit Level 2 PSA are associated with the characterization and quantification of the uncertainties in the modelling of severe accident progression phenomena, MAAP-CANDU parameter uncertainty, containment event tree branch point quantification, containment response under severe accident conditions, and uncertainties on fission product behaviour.

The challenges also include the development of new and emerging methodologies such as:

- Multi-unit PSA, including multi-unit interactions and multi-unit severe accident progression simulation;
- HRA methodology for crediting human actions associated with the deployment of emergency mitigating equipment;
- Methodology for incorporating selected SAMG strategies during multi-unit events;
- Methodology for considering and screening of POS and other radioactive sources.

### 2.2.7. Summary of Session II

There is a consensus among Member States that probabilistic safety goals or criteria for Level 2 PSA are of major importance in the regulatory review for licensing new deployments of NPPs as well as in periodic safety reviews of operating NPPs. These criteria are currently and largely used as a communication measure indicative of the level of safety that a given NPP design can achieve. However, the current status is such that Member States have different approaches to define probabilistic safety goals for Level 2 PSA.

First, regulations in the majority of Member States define criteria or goals either for large early release frequency or for large release frequency, while in other Member States both probabilistic criteria or goals are defined (e.g. Canada). However other terms are also used by some Member States (e.g. specific risk metrics in Switzerland – total risk of activity release in Bq per year or specific risk metric in Canada – small release frequency in Bq of I-131 per reactor-year).

Second, there is a lack of clear, and if possible common, definitions among all Member States of large early release frequency and large release frequency (e.g. Türkiye, Pakistan), due to national considerations related to the definition of terms such as ‘early’ and ‘large’.

Third, some Member States prefer to define Level 2 PSA probabilistic criteria or goals as qualitative rather than quantitative (e.g. France).

Fourth, for some Member States the probabilistic values are set as criteria to comply with as absolute values, while in other Member States probabilistic values are goals to be met considering either a reduced or full scope[4]. This difference of approach between strict criteria or goals has an impact on the how Level 2 PSA results are assessed. Therefore, the importance of having a common understanding and definition of those terms could greatly contribute to the harmonization of the regulatory review process for licensing new NPPs among different

[4] Full scope for probabilistic safety assessment considers internal events, internal hazards and external hazards for both the reactor core and the spent fuel pool for all plant operating states (power operation, low power and shutdown). A reduced scope therefore is commonly considering only a part of the full scope, such as considering internal events only or internal events and some internal hazards and external hazards (see SSG-3 (Rev. 1) [1] and SSG-4 [3]).

Member States by allowing comparison of review results, facilitating independent peer reviews and reinforcing communication and relation among national regulators among Member States.

The technical meeting discussions explored in detail the definition of what is understood as an early radioactive release and of a large radioactive release in different Member States. A common criterion for their definition was pointed out in relation to the acceptable dose limits for individuals in compliance with both the philosophy used for radiation protection purposes recommended by the International Commission on Radiological Protection [25] and with the requirements established in IAEA Safety Standards Series Nos GSR Part 3, Radiation Protection and Safety of Radiation Sources: International Basic Safety Standards [26] and GSR Part 7, Preparedness and Response for a Nuclear or Radiological Emergency [27].

Also, the relationship between large early release frequency and large release frequency was discussed and it was concluded that the former is a subgroup of the latter referring to radioactive releases that might occur in the short-term phase of the severe accident and therefore both provide different key information needed for the regulatory review. During the discussion, the opinion was presented that in practice grouping all large early releases in large releases is acceptable if the large release frequency is lower than the safety goal, especially during the construction stage (this approach was used during Akkuyu NPP Level 2 PSA development).

The second point of major discussion was related to the time value used for defining the limit between an early radioactive release, considered as part of large early release frequency, and a late radioactive release, which would be considered as part of large release frequency. There was agreement in these discussions that the definition of this time value needs to consider the arrangements for the effective implementation of emergency preparedness measures with regard to the acceptable dose limits for individuals (see GSR Part 3 [26] and GSR Part 7 [27]). The example of Finland defining early radioactive releases, which is defined based on emergency preparedness requirements, zoning around the NPP (the distance from populated areas) and most exposed persons was explained in detail. The importance of considering the zone around the NPP in relation to the population such as density, activities or occupation, seasonal fluctuations, infrastructure (e.g. hospitals, schools, hotels) and transport means (e.g. presence of highways or secondary routes for evacuation) as well as environmental aspects such as topography, landscape and climate and meteorological data such as predominant wind directions was highlighted. Additionally, it was emphasised those site aspects of the zone around the NPP limits the effectiveness of the early implementation of measures in relation to pre-arrangements such as distribution of stable iodine tablets, preparation and identification of shelters, communication means to the population (e.g. alarms, telephone messages, radio and television broadcast), education of the population (e.g. regular information to the local communities), and planning of drill exercises related to sheltering and evacuation (e.g. availability of means of transport such as buses and trains). For Finland, the time limit is estimated approximately four hours after information on the start of the emergency situation has been received. The discussion concluded that given the impact of specific site characteristics and Member State emergency arrangements related to the NPP, it is difficult to reach a common consensus on a quantitative limit. However, there was agreement in these discussions that the definition of the term 'early' associated with the large early release frequency risk metric, probabilistic goal or criteria needs to clearly describe the previous discussed aspects, so that comparison among Member States can be performed.

The two main approaches for using probabilistic values, as strict criteria or goals, interested the participants for assessing Level 2 PSA results. The discussions pointed out the advantages and disadvantages of both approaches. For the approach defining strict criteria, the main advantage is related to the regulatory review of Level 2 PSA results allowing a straightforward position for regulatory oversight activities, evaluation of design alternatives, risk management and risk informed decisions. However, this approach has the disadvantage that it could be understood that complying with the strict criteria may reduce the scope of the detailed review process to be performed by the regulatory body and it could hinder the importance of key aspects such as sensitivity, importance and uncertainty analyses. For the other approach, probabilistic values as a goal have the advantage of allowing a flexible regulatory decision making process, where the interest is focused on the validation of the methods and the assumptions used (e.g. assumption for performance shaping factors in HRA, equipment qualification and survivability, flammability limits and combustible gases concentrations) as well as on the uncertainties associated with the result, rather than the final quantitative value. The main disadvantage of this approach is related to the high level of expertise required at the regulatory body which needs to have a sound knowledge of the Level 2 PSA methodology in order to perform the regulatory review.

Another topic of discussion was regarding different approaches among Member States related to the scope of radioactive sources to be considered for the development of Level 2 PSA. Particularly, during the discussion, it was confirmed that Finnish and Russian requirements mean that Level 2 PSA is performed for the spent fuel pool (SFP), i.e. as part of other radioactive sources on site. Finland confirmed that common consideration of a severe accident in the reactor core and the SFP needs to be as realistic as possible. In other Member States, such as Canada, the basis for screening out the SFP from the large early release frequency relies on the time considered to realistically reach fuel uncovered in the SFP which allow sufficient time for external intervention. In relation to the development of Level 2 PSA for the SFP, there was agreement during the discussions that the damage state set at Level 1 PSA for the SFP (i.e. fuel boiling, fuel uncovering, fuel damage) as well as the location of the SFP in the NPP design (i.e. inside or outside a reactor building or a building ensuring a confinement function) may have a major importance.

The topic of importance measures in Level 2 PSA from a regulatory perspective was discussed. It was highlighted that the modelling of the Level 2 PSA can be done by two approaches, by an integrated Level 1–Level 2 PSA model or by separate models in which the plant damage states from Level 1 PSA are used as the starting point for development of Level 2 PSA. In the case of approach 1 during importance analysis it can be noted that importance of Level 1 PSA structures, systems and components tends to dominate the results. So, in order to understand the importance of structures, systems and components performing containment functions, there was agreement during the discussions that it is probably reasonable to start from the plant damage states. However, in this case there are difficulties associated with the phenomenology aspects modelled, which might not be fully reflected in the transition from Level 1 PSA, and the fact that some structures, systems and components are considered in both Level 1 and Level 2 PSA.

The issue of reassessing hazards and their combinations in the development of Level 2 PSA was discussed. The discussion was focused on regulatory requirements to look back into the screening analysis performed as part of Level 1 PSA. Those regulatory requirements, which are in some Member States' regulations, aim at identifying hazards and their combinations which were screened out in Level 1 PSA but could be of interest in Level 2 PSA taking mainly into account the difference in mission times considered between Level 1 PSA and Level 2 PSA (i.e. 24 hours is generally used in the majority of Level 1 PSA and up to several days to a week for Level 2 PSA) and the different systems used (i.e. reactor containment and associated systems are only modelled in Level 2 PSA).

Additionally, during the discussions, technical meeting participants requested the IAEA to have further guidance on the development of Level 2 PSA for different stages of the NPP, in particular at the design stage.

## 2.3. SESSION III: CONSIDERATIONS FOR THE REVISION OF SSG-4

### 2.3.1. Session III description

As mentioned in the previous section, there are several topics of interest for Member States in relation to the development of Level 2 PSA. Given the need for harmonization of practices related to the development and application of Level 2 PSA, Session III was devoted to the discussion of SSG-4 [2] in view of its revision to incorporate current practices and the latest improvements in relation to the Level 2 PSA. Three presentations were prepared presenting key aspects to be considered for the review of SSG-4 [2]. The abstracts of presentations are included in Sections 2.3.2–2.3.4 and the summary of the session's discussions is presented in Section 2.3.5.

It is important to note that the suggestions from the participants in the technical meeting captured in the following sub-sections have been examined in detail and implemented, as appropriate, in the preparation of SSG-4 (Rev. 1) [3], which is ongoing at the time of preparing this publication.

### 2.3.2. Armenia's critical aspects for the review of SSG-4

This presentation provided the background of revision of IAEA safety standards initiated after the accident at the Fukushima Daiichi NPP. The focus of the presentation was hazard combination, source(s) of the radiation to be considered in Level 2 PSA (e.g. spent fuel pool, multi-sources) and severe accident management guidelines (e.g. considerations related to use non-permanent equipment for electrical power supply and cooling capability and the related human actions).

The presentation provided some relevant aspects in relation to SSG-4 [2]:

- SSG-4 [2] provides limited recommendations for external hazards and does not address combinations of the hazards. The presentation provided the approach used in Armenia to address the combination of hazards.
- Despite the fact that SSG-4 [2] mentions the need to consider the potential release from other sources of radioactivity at the plant, such as irradiated fuel and stored radioactive waste, it does not provide recommendations on this topic. The presentation underlined the importance to supplement SSG-4 [2] with recommendations on the SFP considerations including two conceptual design solutions — SFP inside and outside containment — as well as the potential for accidents involving multiple reactor units and SFPs concurrently.
- Regarding severe accident management, SSG-4 [2] uses the term 'beyond design basis accidents', meanwhile, after the publication of SSR-2/1 (Rev. 1) [5], the term 'design extension conditions' was introduced and replaced 'beyond design basis accidents'. The presentation underlined the importance to supplement SSG-4 [2] with recommendations regarding non-permanent equipment (e.g. for electrical power supply), HRA in the context of using non-permanent during severe accident situations, and Level 2 PSA applications in general.

The presentation provided some topics for further discussion, such as how the human resources need to be modelled in the multi-source scenario, how the large early release frequency or the large release frequency could include the SFP, and how to deal with shared equipment.

### 2.3.3. Russian Federation's critical aspects for the review of SSG-4

This presentation provided both general and specific remarks on SSG-4 [2] aimed to support the revision of the Safety Guide. Several aspects of SSG-4 [2] to be considered were discussed in the presentation. The most important remarks are listed below.

- SSG-4 [2] focuses only on fuel in the reactor core. However, releases from fuel in the SFP might have higher frequency and more severe consequences than from the fuel in the reactor.
- SSG-4 [2] does not consider stored radioactive waste in Level 2 PSA. However, in many Member States it is required to include releases from radioactive waste in the scope of Level 2 PSA.
- Numerous undefined terms are used in SSG-4 [2], such as release category and source term, early and large releases, and mission time in Level 2 PSA.
- Section 2.1 of SSG-4 [2], "Definition of the objectives of Level 2 PSA" does not provide any information on recommended or practiced probabilistic safety goals or criteria associated with Level 2 PSA.
- Section 3 of SSG-4 [2], "Identification of design aspects important to severe accidents and acquisition of information" does not contain any recommendations related to the information needed for Level 2 PSA for radioactive sources other than fuel in the reactor, other hazards than internal events and other modes of operation than operation at power.
- Section 4 of SSG-4 [2], "Interface with Level 1 PSA: grouping of sequences" does not provide useful recommendations. In particular, recommendations on the following important aspects are missed; treatment of recoveries in the Level 1–Level 2 interface model, enhancement of the Level 1 PSA model to add features needed for Level 2 PSA, attributes important for Level 2 PSA for shutdown states, consideration of mission time in the Level 1 and Level 2 interface model.
- Section 5 of SSG-4 [2], "Accident progression and containment analysis" does not indicate which end states are assessed in the containment event tree. Containment event tree questions are presented in the form suitable for the 'EVNTRE' code [28]. According to expert opinion, this code is rarely used now as it does not allow an integrated model to be developed. No information is provided on containment event tree structure and nodal questions used in more common software (e.g. RiskSpectrum, Cafta, Saphire, FinPSA, NUPRA). Also there are no recommendations on how hazards, various radioactive sources and operational states are represented in containment event tree models.
- There is very limited mention of containment event tree quantification (only in para. 5.26). Further recommendations are needed in relation to containment event tree quantification, dealing with the size of the model, the treatment of highly probable 'success' branches, and the possibility to use three or more branches in the containment event tree logic.
- Annex III of SSG-4 [2], "Sample outline of documentation" does not have a section related to system models needed for Level 2 PSA (e.g. spray system, containment isolation, systems, important for severe accidents progression and post-melt behaviour (e.g. core catcher, ex-vessel flooding)). The spent fuel pool, shutdown plant operating state, and the development of Level 2 PSA for hazards are not reflected in the structure.

The following updates for SSG-4 [2] were suggested:

- Provide recommendations related to the development of Level 2 PSA for the fuel in the SFP for several cases (SFP inside containment and outside containment, combined releases from SFP and reactor core).
- Provide recommendations related to the development of Level 2 PSA for radioactive waste stored on the site.
- Provide separate sections dedicated to Level 2 PSA for the SFP and other sources of radioactivity, for multi-unit and multi sources Level 2 PSA.
- Add section terms and definitions and define terms such as large, early, mission time for Level 2 PSA, release category, source term.
- Revise Figure 1 and items (3) and (4) of para. 1.7 to give clear and consistent picture of the process.
- Combine Sections 5 and 6 in order to have complete picture of the end points of containment event trees.
- Provide illustrative examples of containment event trees in graphical form.
- Add a section on "Level 2 PSA probabilistic safety goals or criteria".
- Provide recommendations related to the information needed for Level 2 PSA for other sources of radioactivity, for shutdown plant operating state and for external hazards and internal hazards.
- Provide recommendations on how end states in the containment event tree are defined.
- Provide recommendations for containment event tree construction using integrated Level 1 and Level 2 PSA models.
- Provide examples of containment event trees developed using integrated models.
- Provide specific recommendations on specific features of containment event trees for plant damage states developed for hazards and shutdown states (if any).
- Add a separate section on "Containment event tree quantification" (as it is proposed in SSG-3).
- Comprehensively review and update proposed outline in Annex III (SFP, shutdown plant operating state, and hazards PSA need to be reflected in the outline) and to limit the outline only to section level; remove subsections in proposed outline.
- Present specific information in separate Annexes:
  - Approaches for assessment of reliability of passive systems in Level 2 PSA (e.g. core catcher, passive autocatalytic recombiners);
  - Approaches for assessment of certain important phenomena (e.g. grip rupture, direct containment heating, detonation/deflagration);
  - Illustrations for bridge trees and containment event trees;
  - Information on development and use of Level 2 PSA at design stage.

This presentation also provided a further set of specific comments mainly following the general comments.

### 2.3.4. Switzerland's critical aspects for the review of SSG-4

The experience of the development process, the techniques used, and insights gained from the latest integrated, full scope, multistate Level 2 PSA analysis conducted at the Leibstadt Nuclear Power Plant (KKL) was provided in this presentation.

International guidelines and standards such as SSG-4 [2] and Refs [29, 30] have been applied in the development of Level 2 PSA.

The project considered the following risk metrics, as defined by Switzerland:

- Large early release frequency (LERF) – the large early release frequency is the expected number of events per calendar year with a release of more than 2.0 x $10^{15}$ Bq of Iodine-131 per calendar year within the first 10 hours after core damage.
- LRF – the large release frequency is expected number of events per calendar year with a release of more than 2.0 x·$10^{14}$ Bq of Caesium-137 per calendar year.
- Specific risk metrics in Switzerland – The total risk of activity release provided in Bq per year, serves as an indicator for the anticipated overall release of radioactive material following core damage within a given calendar year. It is computed by multiplying the frequency of each release category by its corresponding source term, the CDF, and then summing up these products.

Based on this experience in Switzerland, the following aspects were added to the previously identified key points for the revision of SSG-4 [2]:

- Treatment of uncertainties (especially phenomenological) (aleatory/epistemic);
- Nodal probabilities;
- Suggestions related to severe accident computer codes and models;
- Coupling requirements;
- Deflagration-to-detonation transition, flame accelerations;
- Definitions of large early release frequency and large release frequency and safety goals, including the definition of terms such as 'early' and 'large';
- Result presentations;
- Suggestions on corium coolability;
- Importance analysis based on Level 2 measures;
- Sizing of emergency zones (Level 3);
- Suggestions related to the modelling of additional safety features considered for design extension conditions in Level 2 PSA need to include the filtered venting system.

Such aspects may be of a lower level than SSG-4 [2], but they may affect results of the Level 2 PSA. Accordingly, discussion on these aspects needs to be encouraged through the expert community.

### 2.3.5. Summary of Session III

In addition to suggestions from the presentations, during the discussions the need to provide recommendations for the definition of Level 2 PSA risk metrics, goals or criteria was highlighted. This suggestion was considered in the preparation of SSG-4 (Rev. 1) [3], ongoing at the time of preparing this publication. It was confirmed that a section exists on safety goals and criteria in SSG-3 (Rev. 1) [1], which provides general information for Level 1, Level 2, and even Level 3 PSA, but without elaborating on the definition of particular Level 2 PSA risk metrics, goals or criteria such as the large early release frequency and the large release frequency. Also, it was confirmed that quantitative values are difficult to present in SSG-4 [2] since they could challenge the consensus process for Safety Guides. Therefore, the revised version included the definition of the Level 2 PSA probabilistic safety goals as well as some examples in the Annexes.

During the discussions it was also requested to include recommendations related to the integral and separate approaches for the development of Level 2 PSA including their advantages and disadvantages. In relation to this aspect, it was commented that the methodology proposed in SSG-4 [2] starts from a Level 1 PSA, as separate approach. This approach might not be applicable to advanced NPPs, including SMRs, since the Level 1 PSA and its associated risk metrics might not be relevant for them. These advanced NPPs propose new design safety features aiming primarily at reinforcing the achievement and maintain the fundamental safety functions. Many of these advanced NPP designs have a smaller reactor thermal power that facilitates the residual heat removal by using passive safety systems combined with natural circulation, which results in achieving rather low frequencies of core damage [31]. In addition, these advanced NPP designs reinforce the confinement of radioactive material by including specific design features such as compact designs and new materials for the fuel matrix which differs with current water cooled reactors [31]. The revised version considered these suggestions and expanded better the recommendations related to the integral approach for both current NPPs and new designs based on advanced NPPs.

In addition, during the discussion it was requested to explicitly define the methodology used as technology neutral. In that regard, the SSG-4 (Rev. 1) [3] aims at presenting recommendations as technology inclusive rather than technology neutral. The understanding of technology neutral term will apply only to a general PSA methodology where no specifics related to reactor technology are presented. In the revised version, the technology inclusive methodology term is proposed since the recommendations there recognize specific aspects related only to NPPs, such as reactor core, reactor containment, SFP and release categories. However, to complement the technology inclusive methodology, it is understood that the phenomena associated with severe accident are technology and design specific.

Finally, the discussions also commented on the request whether or not to include considerations related to the development of multi-unit Level 2 PSA. For this topic, no general consensus was reached since there is no general experience in Member States. However, most of the technical meeting participants agreed that single unit Level 2 PSA needs to consider potential interactions and dependencies between units at the same site. The revised version has taken this suggestion into consideration and allows for the flexibility related to multi-unit Level 2 PSA.

## 2.4. SESSION IV: RESEARCH AND DEVELOPMENT PERSPECTIVE

### 2.4.1. Session IV description

Session IV was devoted to an overview of the current practices in Member States related to research and development on Level 2 PSA. During the session, the presentations covered different topics such as the development and use of computational tools and calculation codes in support of Level 2 PSA, consideration of Level 2 PSA for facilities with SMRs and research reactors, an overview of the new ASME standard on Level 2 PSA, definition of the list of design extension conditions, and overviews of the experience in development of Level 2 PSA in Member States. Participants from Argentina, Brazil, Finland, France, Japan, Republic of Korea, Russian Federation, and United States of America presented their experiences and approaches in Level 2 PSA. The abstracts of presentations are presented in Section 2.4.2–2.4.12. The discussion summary of the section is presented in Section 2.4.13.

### 2.4.2. In-vessel phase of accidents with core meltdown for Level 2 probabilistic safety assessment, Russian Federation

Severe accidents involve complex physicochemical and radiological phenomena. The accident phases associated with these phenomena are typically categorized into two groups:

(1) In-vessel phase, which encompasses core heat-up, fuel degradation, and material relocation expected to occur inside the reactor vessel up to the failure of the reactor vessel. It also involves the subsequent release of molten corium into the containment building.

(2) Ex-vessel phase, which encompasses thermal and chemical interactions between core debris and containment structures, and containment behaviour, which includes transport of radioactive substances inside of the containment.

In this presentation the basic information of the computer code SOCRAT structure was presented and the functionality of the separate modules of the code was described. SOCRAT is Russia-developed severe accident code SOCRAT/B1, which is widely used for design modelling of accidents with severe core damage.

The advantages of modern codes include:

- Use of nodalizations for the primary and secondary circuits similar to those used for analyses in thermohydraulic codes;
- Modelling of normal operation systems and safety systems;
- Establishment of a steady state condition;
- Accident analysis from the initiating event;
- Adjustment following the analysis results with the use of thermohydraulic codes.

The structure of SOCRAT modules for modelling of severe accidents is as follows:

SOCRAT/B1 is a modern tool for severe accident analysis including modules:

- RATEG – primary and secondary thermal hydraulics.
- SVECHA – core degradation.
- HEFEST – thermal physics of corium and thermal mechanics of reactor pressure vessel (RPV).
- ANGAR, KUPOL – thermal hydraulics in containment considering:
  - Damage and melting of the core fuel elements;

- Generation and transport of non-condensable gases and hydrogen;
- Core barrel and reactor vessel damage;
- Release of mass and energy from the reactor vessel.

SOCRAT/B3 includes:

- ANGAR, KUPOL – thermal hydraulics in containment;
- TOCHKA – point neutron kinetics;
- BONUS – fission products accumulation in fuel;
- RELEASE – fission products accumulation in fuel;
- MFPR_MELT – fission products release from the corium;
- GAPREL – fission products release from the gas gap;
- PROFIT – fission products behaviour in the primary circuit;
- CONTFP – fission products behaviour in containment volumes;
- RACHIM – chemical reactions, activity, heat generation by fission products.

The severe accident phenomena modelled by the combinations of codes SOCRAT B1 + B3 considers:

- Reactor unit thermohydraulics;
- Core damage;
- FP (fission product) release from fuel;
- FP release from corium;
- FP transfer into reactor containment
- Melt moving;
- Melt pool;
- Fuel coolant interaction;
- RPV damage.

OKB "Gidropress" JSC has acquired extensive experience in analysis of severe accidents for the project justification of safety of existing and prospective NPPs with VVER designs. The base code for the analysis of severe accidents in Russia is the SOCRAT/B1 code. However, for the analysis of the release and transfer of fission products, a new version of the SOCRAT/B3 code could be used. As of now, in the framework of the research, the cooperation scheme has been tested by OKB "Gidropress" JSC by performing calculations of severe accidents with SOCRAT/B3 to support Level 2 PSA. It is planned to apply this test scheme in the future.

### 2.4.3. Simulation based approach to Level 2 probabilistic safety assessment, Finland

The timings of events are more important in Level 2 PSA than Level 1 PSA, because physical phenomena and the recovery of safety functions need to be modelled. For example, if a water cooled reactor core is reflooded during a critical time window, significant amounts of hydrogen are produced, possibly leading to a hydrogen explosion. Traditional fault tree-based modelling is not very well suited to modelling such time dependencies and phenomena. In addition, the set of possible accident conditions is so large that it cannot properly be captured in a binary model.

VTT has developed a simulation-based Level 2 module to their FinPSA software. The Level 2 module combines event trees with script-based modelling. The model includes a script file for each event tree header. In the script files, functions are defined to calculate conditional probabilities of event tree branches as well as source terms for accident sequences. The script files offer lots of flexibility for modelling, and it is possible to include timings of events explicitly in the model with uncertainty distributions. Dynamic dependencies related to severe accident phenomena can be modelled in the scripts. The model is also not restricted to binary logic. A branching point can include more than two branches, continuous variables can be used, and various/different conditions and accident timing scenarios can be incorporated in the scripts.

Uncertainty distributions can be defined for all variables of the model, and the model is solved by Monte Carlo simulations. The result is a set of simulation results for each accident sequence. Then, for raw simulation data, statistical analyses are performed to calculate mean results and uncertainty distributions. The tool also includes a risk integrator that combines simulation results from multiple containment event trees.

The Level 2 model can be integrated with the Level 1 model built using a traditional PSA approach with event trees and linked fault trees. The Level 2 tool can read Level 1 accident sequences and minimal cut set results. Level 1 information can be utilised in Level 2 modelling, for example to calculate the core cooling recovery probability. The contributions of the most important Level 1 sequences, basic events and initiating events can also be seen in Level 2 results.

FinPSA Level 2 is used for Level 2 PSA of the Olkiluoto NPP units. In addition, VTT has conducted limited case studies on some severe accident phenomena and research on script-based modelling techniques. For example, Level 2 PSA of a boiling water reactor (BWR) plant was studied by performing deterministic simulations using MELCOR and utilizing the results of those in the Level 2 PSA modelling. It was studied how the recovery time of emergency core cooling and depressurization time affect later accident progression, and those time dependencies were implemented in the scripts.

The BWR study particularly focused on ex-vessel steam explosions. The probability of containment failure due to an ex-vessel steam explosion was estimated based on the probability distributions of the pressure impulse and containment strength in different scenarios. Several dependencies related to the probability of an explosion and containment failure were modelled explicitly in the scripts. The pressure impulse distribution depends on the pressure and the amount of core melt ejected to the lower drywell. Whether an explosion can be triggered at all depends on the lower drywell flooding time and core meltdown timings.

A time dependent source term model influenced by the so-called XSOR method was also implemented in the simulation scripts. The release mechanisms that were modelled were early reactor coolant system release, late reactor coolant system release (e.g. re-vaporization of deposited fission products) and ex-vessel debris release. The release start and end times were determined for each release mechanism based on e.g. core meltdown timings and vessel failure time. The release time intervals were divided into 10 subintervals, and discrete point releases were calculated at these subintervals. Finally, the discrete point releases were summed to calculate the total releases. This approach makes it possible to perform a large number of simulations to calculate uncertainty distributions for source term variables.

The case study demonstrated how FinPSA Level 2 can be used. Results from deterministic analyses, such as different timings, can explicitly be incorporated into the scripts, and distributions can be specified for uncertain variables. A source term model can also be integrated to the simulation-based PSA model. The tool however enables several different modelling approaches. It is also possible to develop simplified physical models in the simulation scripts of FinPSA, instead of using external deterministic simulation software like in the case study. The programming-based modelling gives users a lot of freedom in the selection of the modelling approach and the level of modelling detail, as shown in Figure 2.

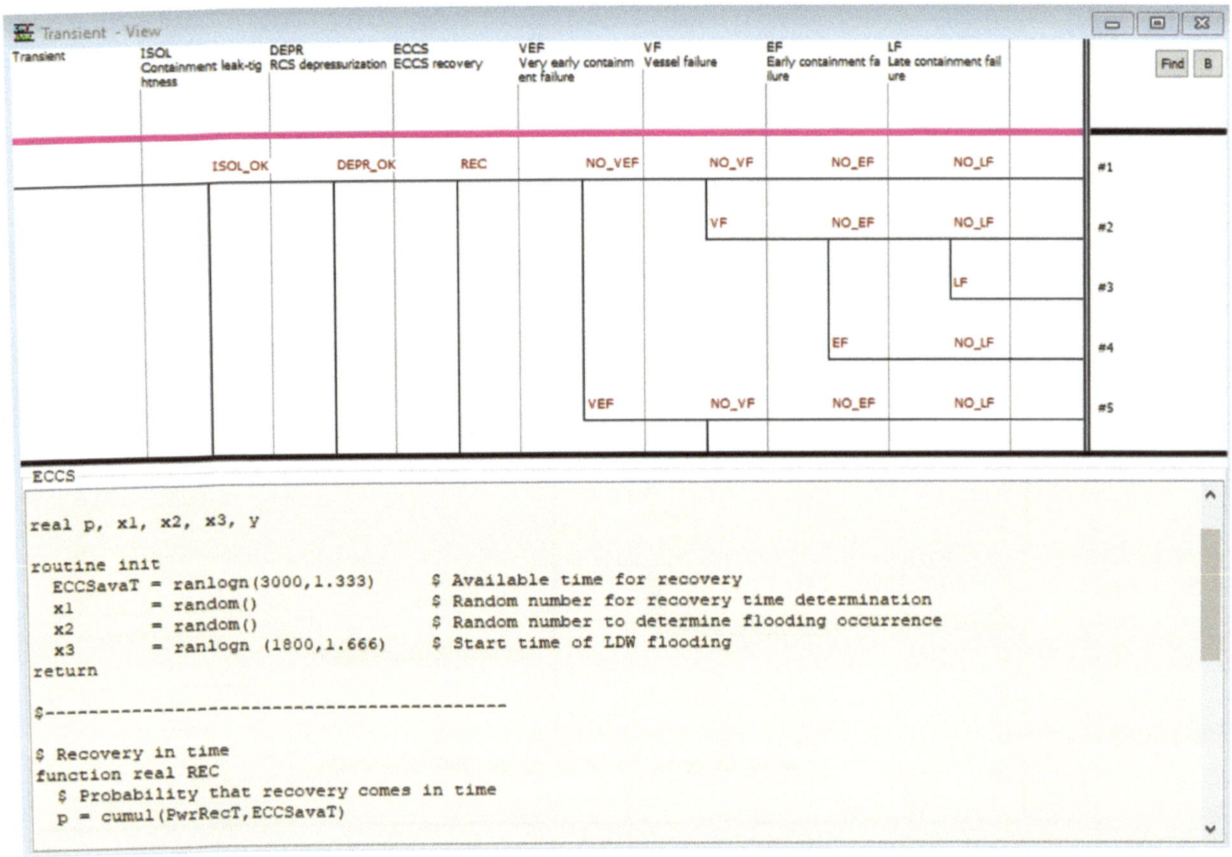


*FIG. 2. Simulation-based approach in Level 2 PSA (courtesy of T. Tyrväinen, VTT Technical Research Centre of Finland, Finland).*

### 2.4.4. CAREM-25: Level 2 probabilistic safety assessment activities in the context of SSG-4 application, Argentina

#### *2.4.4.1. CAREM-25 SMR prototype*

Central Argentina de Elementos Modulares (CAREM-25) is a national project focused on developing SMRs using LWR technology. Coordinated by the CNEA in collaboration with prominent nuclear companies in Argentina, CAREM-25 aims to design and construct innovative, economically competitive, and highly safe NPPs. The deployment of CAREM-25 serves as a prototype to validate the innovations intended for the future commercial version of CAREM.

CAREM-25 features design elements such as an integrated primary coolant system, self-pressurization, core cooling via natural circulation, and in-vessel hydraulic control rod drive mechanisms. Notably, its passive safety systems activate during a 36-hour grace period, storing released energy within the containment building.

#### *2.4.4.2. Design internalization of the defence in depth concept*

In the design of CAREM-25, the concept of defence in depth has been integrated since the project's inception. This approach serves as the fundamental framework for classifying structures, systems, and components important to safety. The applied defence in depth concept aligns with the proposal by the Western European Nuclear Regulators Association

(WENRA) [32] and emphasizes clarity regarding multiple failure events, severe accidents, and independence between safety levels. The adopted approach in CAREM is schematically presented in Fig. 3:

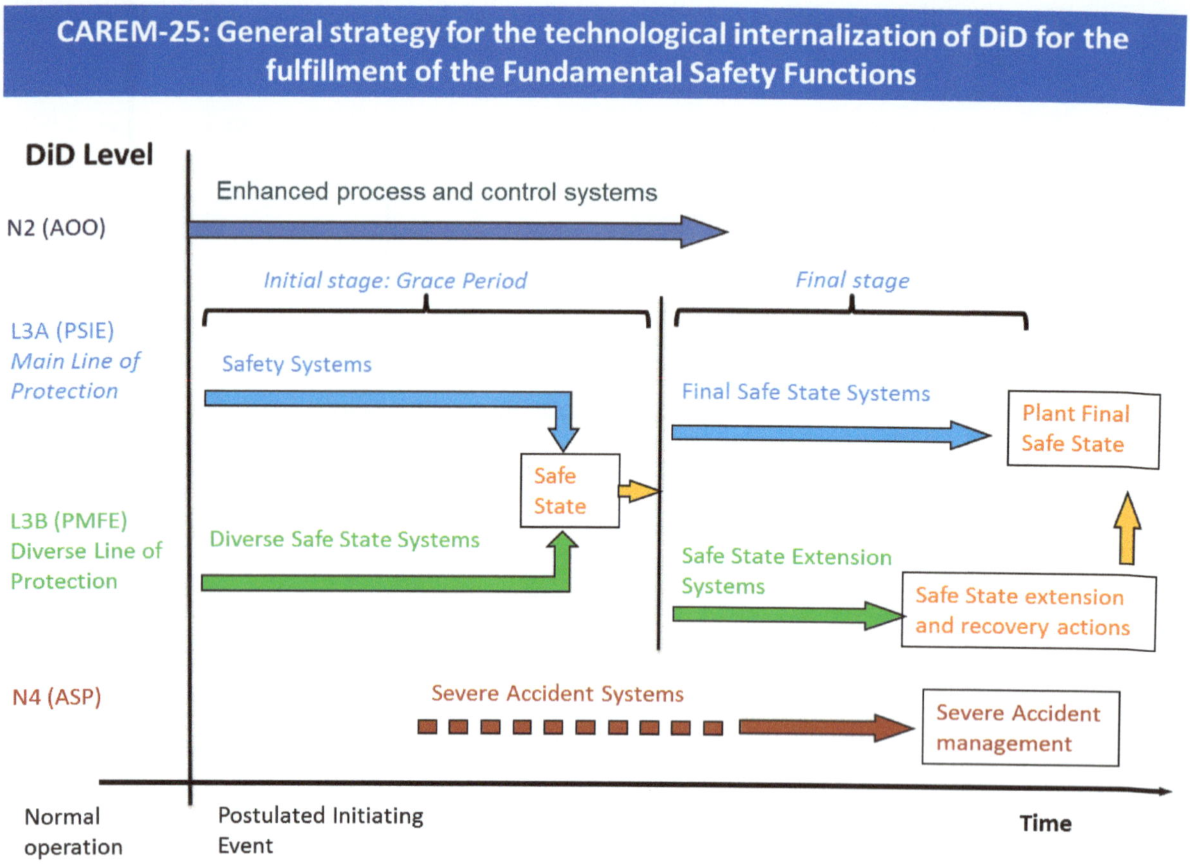


*FIG. 3. Defence in depth internalization in CAREM-25 (courtesy of M. Gimenez, National Atomic Energy Commission, Argentina).*

In severe accident conditions (level 4 of defence in depth), the objective is the mitigation of postulated core melt accidents to terminate the progression of core damage once it has started; to preserve the integrity of the containment as long as possible; to minimize releases of radioactive material and to achieve a long term stable state, by means of severe accident mitigation systems. Consequently, the following defence in depth level 4 safety functions were defined:

- Limit the hydrogen concentration in the containment to reduce the possibility of hydrogen deflagration and detonation;
- Remove decay heat generated in the molten material relocated into the reactor pressure vessel lower head to preserve its integrity;
- Enhance iodine retention in suppression pool water to reduce the iodine activity released in case of containment failure;
- Limit the containment pressure to preserve its integrity, in order to limit possible radioactive releases.

In addition to this, a set of defence in depth level 4 monitoring safety functions were defined, in order to evaluate the performance of severe accident mitigations systems and to assess the different plant damage conditions to be considered for the development of severe accident management guidelines.

#### 2.4.4.3. *Systems important to safety related to severe accident mitigation*

The most relevant of these systems are the following:

- **Severe accident mitigation systems (defence in depth level 4):**
  - Hydrogen control system;
  - In-vessel melt retention;
  - Iodine suppression pool retention (pH increase);
  - Containment venting.
- **Extension of plant safe state systems (defence in depth sublevel 3b):**
  - Reactor pressure vessel water injection system by external means;
  - Passive reactor heat removal system pool water injection by external means;
  - Suppression pool cooling (heat exchanger with external water supply);
  - Passive reactor heat removal system chamber cooling (heat exchanger with external water supply).

#### 2.4.4.4. *Severe accident deterministic analysis*

Analytical and computational analyses are being carried out to determine how different accident sequences could challenge critical safety functions and how the different fission product barrier could be compromised or damaged.

An integral MELCOR 1.8.6 model was developed to simulate the whole plant dynamics. In this model primary system, secondary system, containment compartments and defence in depth Level 3 and Level 4 systems important to safety were represented. This model has been used to study severe accident phenomena and containment dynamic behaviour.

#### 2.4.4.5. *Severe accident probabilistic analysis*

Preliminary probabilistic severe accidents analyses are under development. This includes a preliminary Level 1–Level 2 PSA interface, taking the current available Level 1 PSA as a basis; a simplified containment event tree model, using the RiskSpectrum code; and a general definition of attributes as part of the PSA Level 2–Level 3 interface, for the assessment of the corresponding release categories.

### 2.4.5. Proposed Level 2 probabilistic safety assessment methodology for a vSMR, Brazil

SMRs are smaller than reactors in conventional NPPs and are typically designed to produce up to 300 MW(e)[5]. Very small reactors (vSMR)[6] (about 10 to 50 MW(e)) are a category of SMR.

This presentation provided the development and application of a Level 2 PSA methodology for a generic pressurized water vSMR with 45 MW(th) and 11 MW(e) during low power and shutdown phases of plant operation.

The proposed methodology follows the recommendations in SSG-4 [2] and also is consistent with Refs [33-35] and with specific features associated with the accident phenomenology and reactor type.

---

[5] As defined by the SMR Regulators' Forum

[6] Some Member States define these SMRs as microreactors.

The representative severe accident sequences were analysed with the MELCOR code. Containment event trees were used to analyse and represent accident progressions of the plant damage states. The progression of the plant damage states was modelled through an event tree using the computer aided fault tree analysis, CAFTA. The shutdown model was quantified using PRAQuant version 5.2.

The methodology for performing a Level 2 PSA was divided into six main steps:

Step 1: Selection of initiating events.

Step 2: Plant damage state grouping.

- Plant damage states grouping criteria:
    1. Analysis of common cause failures (e.g. of valves, heat exchangers, pumps, reactor protection system components) in redundant systems of interest;
    2. Immediate effect (e.g. loss of coolant circulation in the reactor core or SFP, loss of circulation of cooling water from the residual heat removal system, loss of cooling in the SFP or loss of SFP heat sink);
    3. Availability of safety systems (e.g. performance of heating, ventilating and air conditioner systems of the reactor building, containment isolation systems, exhaust systems, passive autocatalytic recombiner (if any) and performance of containment mitigation systems);
    4. Affected coolant circulation region;
    5. Phases of operation of plant (shown in Table 1).

TABLE 1: OPERATION PHASES PRESENT IN LOW POWER AND SHUTDOWN MODE

| Phase | Description | Fuel localization | Status of the upper part of the reactor vessel |
|---|---|---|---|
| I | Reactor core cooldown | RPV | On (closed) |
| II | Fuel handling/offload | RPV and SFP | Off (open) |
| III | Unload of all fuel | SFP | Off |
| IV | Fuel handling/reload | RPV and SFP | Off |
| V | Reactor restart | RPV | On |

- Circulation regions of the coolant grouping criteria:
    - The circulation regions of the coolant are divided as:
        - Region 1: region through which the reactor coolant flows;
        - Region 1*: region through which the water from the cooling pumps flows;
        - Region 2: region through which the cooling water from the heat exchanger flows;
        - Region 3: region through which the SFP cooling flows.
    - Grouping of the operation phases criteria:
        - Location of the fuel (e.g. in the reactor core or in the spent fuel pool);
        - Duration of the phase;
        - Reactor vessel (e.g. closed or open).

As a result, four groups of phases were formed.

- First plant damage states grouping:

- The 50 accidents sequences were initially grouped into 12 plant damage states taking into considerations the similarity in the:
  - Operation phases;
  - Circulation regions of the coolant;
  - Consequence severity of individual accidents;
  - Frequency of occurrence category.
- Second plant damage states grouping:
  - The second grouping into four plant damage states took into account the: severity of the accident sequence category associated with each plant damage states, frequency of occurrence category associated with the similar operation phase of the plant, and the circulation region of the coolant. And then, the plant damage states were grouped into four plant damage states as shown in Table 2.

TABLE 2: PLANT DAMAGE STATES SET REDUCED INTO 4 GROUPS

| PDS | CET heading | Frequency of occurrence | Normalized frequency (%) |
|---|---|---|---|
| 1 | Loss of coolant circulation in SFP (Phases II & IV) | 2.54E-06 | 5.60 |
| 2 | Loss of coolant circulation in SFP (Phase III) | 2.55E-05 | 56.02 |
| 3 | Loss of coolant circulation in reactor core (Phases I & IV) | 1.46E-05 | 32.10 |
| 4 | Loss of coolant circulation in reactor core (Phases II & IV) | 2.85E-06 | 6.28 |

From the frequency of occurrence perspective can be seen that the major contribution resulted from the loss of cooling accident in the SFP (plant damage state 2 phase III).

Step 3: Accident progression modelling using a containment event tree approach.

Step 4: Radioactive release category identification.

Step 5: Source term analyses.

Step 6: Large early release frequency calculation.

From a consequences perspective, the loss of coolant circulation accident in the spent fuel pool with a closed reactor (plant damage state 1 phase II and IV) was the worst potential accident, having lower frequency but higher consequences.

The preliminary results presented summarized the methodology under development since there is no information available regarding the development of Level 2 PSA for reactor designs with an electrical capacity less than 10 MW(e).

#### 2.4.6. Current status of Level 2 probabilistic safety assessment research, the Republic of Korea

This presentation discussed the following:

Multi-unit Level 2 PSA is one of the critical issues because approximately 70% of the plant sites worldwide are multi-unit sites.

Multi-unit risk metrics are defined as:

- Multi-unit CF – containment failures for two or more units;

- Multi-unit CFF – the frequency that there are containment failures for two or more units;
- Multi-unit LERF – the frequency that there are large early releases for two or more units.

Caesium-137 risk analysis:

- The safety regulations in Korea stipulate that the overall frequency of accidents involving the release of more than 1E+14 Bq of radionuclide Cs-137 is to be less than 1.0 x $10^{-6}$ per reactor-year;
- Ways to decrease Cs-137 risk via management of the structures, systems and components, as shown in Figure 4:
  - Level 1 and 2 PSA linked model with supporting software;
  - Comparison of risk importance measure by CDF and Cs-137 risk;
  - Active management for compliance with the Cs-137 1E+14 Bq rule.

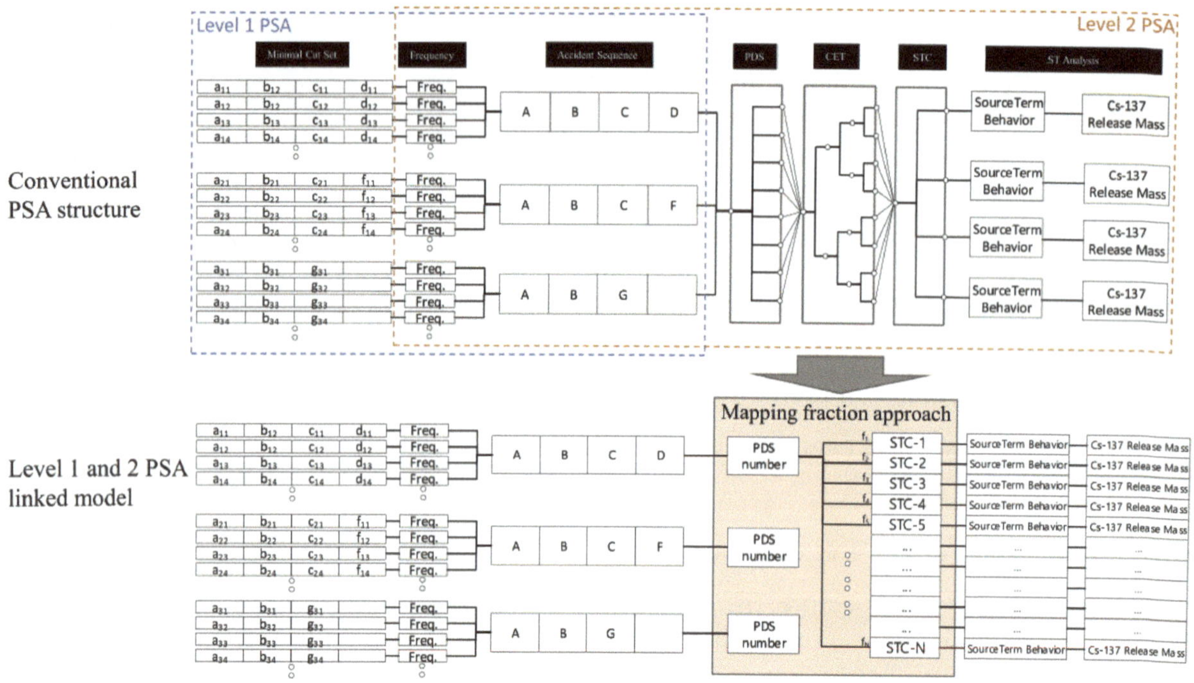


*FIG. 4. Two types of Level 2 PSA modelling (courtesy of J. Cho, Korea Atomic Energy Research Institute, Korea, Republic of).*

SAMG modelling into Level 2 PSA:

- Current Level 2 PSA models consider to some extent the severe accident mitigation strategies employing MACST equipment and severe accident management guidelines: they are limited to specific scenarios or not based on a systematic method.
- Applying severe accident management guidelines contributes to NPP risk reduction.
  - No containment failure probability increased from 37% to 70%;
  - Overall NPP risk significantly decreased by 44%, as shown in Figure 5.

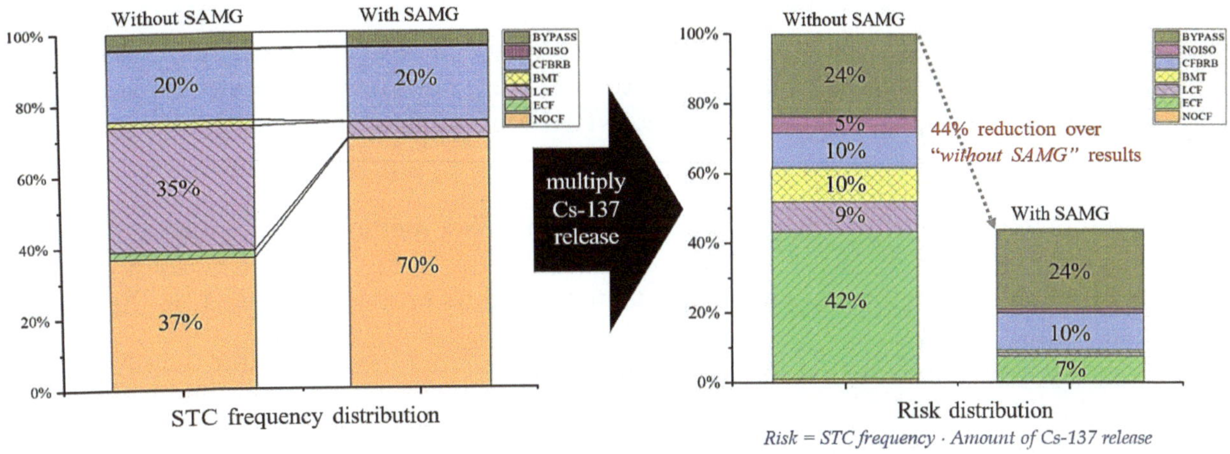


*FIG. 5. Risk frequency distribution (courtesy of J. Cho, Korea Atomic Energy Research Institute, Korea, Republic of).*

Exhaustive source term analysis:

- The conventional grouping method is too conservative. Exhaustive source term analysis is much closer to reality compared to the conventional grouping method, as shown in Figure 6.

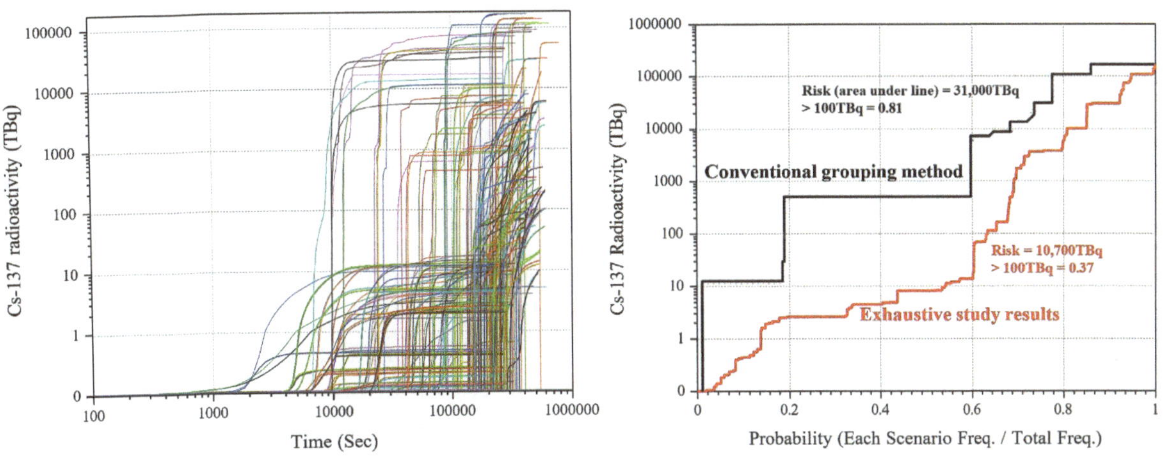


*FIG. 6. Conventional vs exhaustive grouping methods (courtesy of J. Cho, Korea Atomic Energy Research Institute, Korea, Republic of).*

The presentation also discussed expert judgement for severe accident probabilities, the HRA method for supporting Level 2 PSA and a fast severe accident prediction model using a deep learning technique.

#### 2.4.7. Experience in the development and application of Level 2 probabilistic safety assessment for NPPs in NRRC and Japanese industries

The Nuclear Risk Research Center (NRRC) presented their experience in the development and application of seismic Level 1–2 PSA, tsunami Level 1–2 PSA, and related hazard studies.

It is necessary for Level 2 PSA to quantify the branch probabilities of containment event trees, which are values related to not only system failures or loss of functions but also occurrence of phenomena. The phenomenological relationship diagram method has been advanced and applied to quantify probability density functions of molten core concrete interaction, dryout heat flux, etc. In the interface between Level 1 and Level 2 of seismic PSA, NRRC introduced a method for constructing system event trees and phenomenon event trees, respectively, and linking them together. Generally, the final status of safety features is categorized as a choice from success or failure. In addition to these, NRRC introduced the classification of partial success as a medium status.

NRRC presented experiences in developing the tsunami PSA methodology, such as:

- A method for setting the magnitude of a tsunami and its probability based on the tsunami hazard contribution;
- A fragility evaluation method for watertight doors;
- An evaluation method for the fragility of buildings and equipment due to wave force;
- A fragility evaluation method for buildings, structures and equipment by drifting debris;
- An evaluation method of tsunami-specific human error events by applying HRA based on detailed qualitative analysis;
- An evaluation method of fragility of SSCs considering inundation in buildings by tsunami;
- A thermal and chemical effects evaluation method for plant behaviour under inundation of a reactor building by tsunami
- A source term evaluation method considering the inundation of a reactor building by tsunami;
- A success criteria analysis method based on the time dependent tsunami inundation in a reactor building;
- An uncertainty analysis method for tsunami PSA from Level 1 to Level 2 (Figure 7).

Considering drifting debris, a method to determine the fragilities and failure probabilities of target SSCs was developed. Failure probabilities were quantified based on Eulerian-Lagrangian uncertainty analysis of tsunami and drifting debris flow.

An evaluation method for component fragility considering time dependent inundation of buildings by tsunami was introduced. An analysis code 'SHINSUI' for transient inundation propagation into reactor building was developed by NRRC, and the uncertainty of realistic response was quantified based on the tsunami hazard contribution. The code can consider flooding routes, such as doors, hatches and floor openings, air conditioning ducts, etc. Numerical analysis methods of transient free-surface inundation are based on a 'node and junction model' like RELAP, and 'shallow water equations' that are used for compartments with large floor space. Then the timing of the loss of function of pumps and power panels located in each compartment can be calculated according to the tsunami entree route and flow rate (Figure 8).

It was presented that the accident sequence model is handled in three stages. The first step is to model the tsunami scenario event tree based on classification of tsunami impact and plant information, the second is to develop the accident mitigation model based on an internal PSA model and tsunami-specific conditions, and the third is to consider the tsunami-induced fragility. In the seismic PSA, NRRC adopted a direct link model to treat Level 1 and Level 2 event trees together.

A realistic HRA was conducted based on interviews with personnel in an NPP, in accordance with the procedures of the NRRC HRA guide [36]. The improved HRA method made it possible to conduct HRA for emergency operations by multiple personnel in external events, multiple execution tasks, time progression due to interrelationships and factors specific to external events, such as actions to close watertight doors, for which no evaluation method was previously available (Figure 9).

The source term analysis considering thermal and chemical effects in the case of flooding in a reactor building by tsunami was carried out using MAAP5. The interior of the reactor building was divided into several compartments, and the volume and height of each compartment were modelled so that the time dependent inflow of seawater by tsunami was able to be analysed.

NRRC mentioned that they developed Level 1 and Level 2 tsunami PSA methodologies to understand the plant response due to inundation of the site and buildings by tsunamis that overflowed the tsunami protection facilities and equipment. The methods also enable evaluations that reflect physical phenomena related to tsunamis, such as a variety of containment failure modes caused by differences in tsunami height in each accident scenario during severe accidents. The methods will contribute to the quantitative understanding of the risk reduction effect of tsunami protection and mitigation measures, which could not be evaluated by the existing deterministic methods, and to the identification of vulnerabilities to tsunami that could not be found by conservative evaluation.

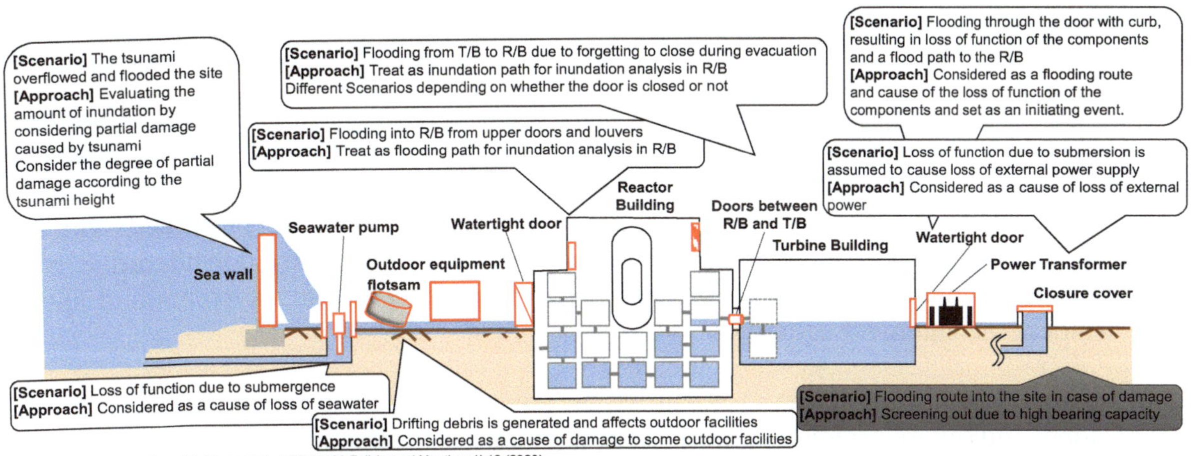


Based on the proceeding of H. Miura, et al., AESJ 2020 Fall Annual Meeting, 1L12 (2020)

*FIG. 7. Modelling of tsunami scenario in NRRC (courtesy of A. Ui, Nuclear Risk Research Center, Japan).*

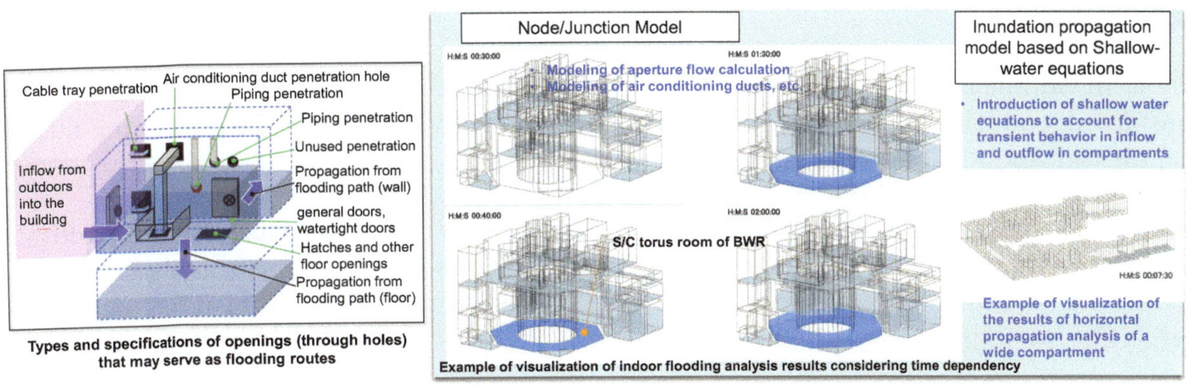


T. Kato, et al., AESJ 2020 Fall Annual Meeting, 1L14 (2020) H. Yamada, et al., AESJ 2020 Fall Annual Meeting, 1L11 (2020)

*FIG. 8. Time dependent flooding analysis in buildings (courtesy of A. Ui, Nuclear Risk Research Center, Japan).*

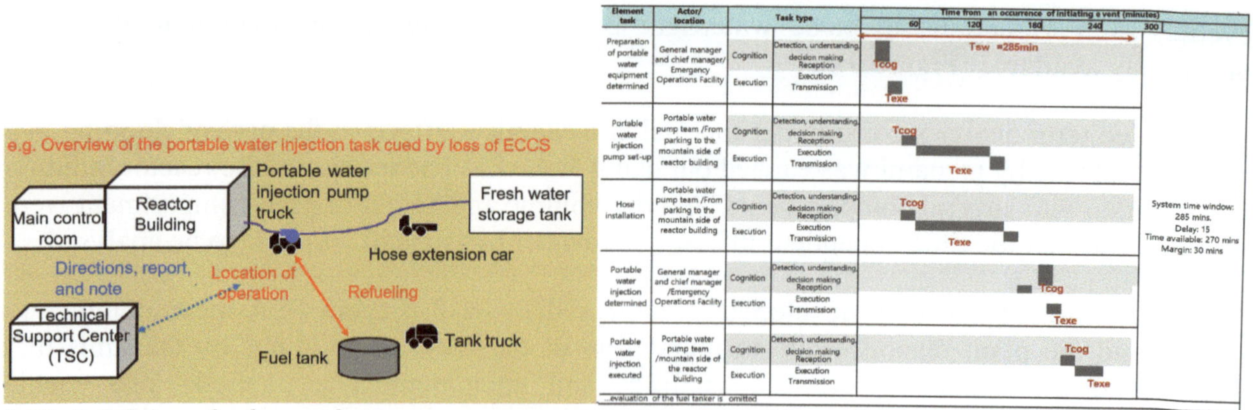


*FIG. 9. HRA method considering non-permanent equipment and loss of watertight door closing (courtesy of A. Ui, Nuclear Risk Research Center, Japan).*

### 2.4.8. Using a Level 1 and Level 2 probabilistic risk assessment[7] model for the development of an ultimate list of beyond design basis accidents, Russian Federation

The methodology for developing an ultimate (fully completed) list of beyond design basis accident has been developed, using a Level 1 and Level 2 probabilistic risk assessment (PRA) model and results. The task of developing such an ultimate beyond design basis accident list is defined by Ref. [12].

According to Ref. [12], the ultimate list of beyond design basis accidents (i.e. including severe accidents) has to be presented in the final safety analysis report. This list has to include representative scenarios for developing and prioritization the management strategy for these accidents. The representativeness of these scenarios is determined by consideration of the different levels of plant state conditions and the availability of safety systems and specific severe accident management equipment.

The ultimate beyond design basis accident list acts as an input for performing beyond design basis accident and severe accident deterministic calculations to support plant safety assessment, using a best estimate approach and uncertainty assessment. The results of these analyses are necessary for the development of beyond design basis accident and severe accident management guidelines and for emergency planning for the protection of plant personnel and the local population.

The term 'ultimate list' refers to how this list has to be fully specific to the NPP power unit considered, and has to present the efficiency of all protective measures for preventing nuclear fuel melting and loss of containment.

In Russia since 2018 a completely new safety guideline with regard to developing of an ultimate list of beyond design basis accident is applicable [37]. This guideline presents the method for developing an ultimate beyond design basis accident list to meet the high-level requirements in Ref. [12].

Following the Ref. [37] methodology, several specific stages are defined for the development of the beyond design basis accident list. The overall procedure is presented in Fig. 10.

[7] The term Probabilistic Risk Assessment is equivalent to Probabilistic Safety Assessment.

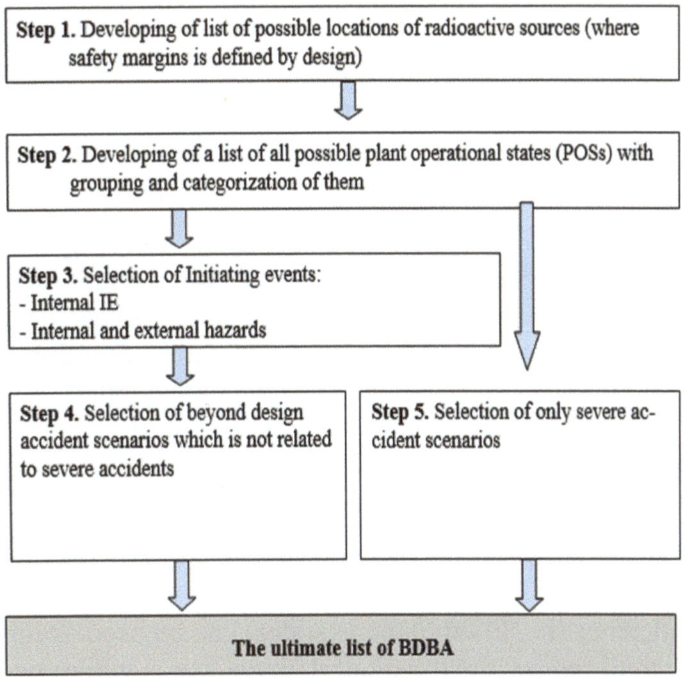


*FIG. 10. Overview of the RB-150-18 [37] procedure (courtesy of N. Stebenev, JSC Atomproekt, Russian Federation).*

The PRA result and models can effectively supplement this RB-150-18 [37] procedure with the following steps.

Step 1: Elaborating of all possible locations of radioactive sources (fissile materials, radioactive substances and rad. waste).

- The up to-date PRA model includes by default the nuclear fuel in the reactor core and in spent fuel pool. Some of PRA models include also the pilot assessment of fresh fuel storage facility.
- The reactor core and spent fuel pool has the determinative input in PRA results and to the fullness of developing of beyond design basis accident and severe accident list. Other radioactive sources can be assessed by qualitative expert judgment.

Step 2: Developing the list of plant operating states.

- Good quality PRA model has to consider all possible plant operating states due to modern requirements to PRA, including changings in safety systems and non-safety systems configuration in each plant operating state. With regard to the fuel in the reactor core and spent fuel pool all plant operating states is considered in PRA.
- Small review is needed to be sure what during categorization and grouping of plant operating states any issue important to this methodology was not forgot.

Step 3: Developing of possible initiating event list.

- The developing of initiating event list in this methodology is fully complemented to the common PRA approach for categorization and grouping of initiating events.
- Small review is needed for PRA results with regard to check if any potential initiating event was screened out of PRA due to low frequency of its occurrence or due other reasons.

Step 4: Developing of beyond design basis accident list, developing the list of safety function and corresponding design measures for the performing of these functions for each plant operating states, developing matrix table: safety functions/design measures VS Initiating events groups.

- There is possibility to use level 1 PRA event trees for determination of accident sequences which has the end state at the level just one step (one failure or human error) before reaching the conditions with nuclear fuel damage.

Step 5: Developing of severe accident list: Consideration the 4 physical barriers and several levels of their degradation; The developing of list of physical phenomena whose compose the danger to physical safety barriers; Developing of graded levels of physical barriers and corresponding states of safety functions; Developing of safety function list for safety function which condition are important to the severe accident management strategy; Developing of generic event trees, with presentations of accident evolution from initiating event to the different level of plant severe state, with regard to loss of considered safety functions and severe accident management measures.

- There is possibility to use level 2 PRA event trees for sufficient performance of whole step 5.
- PRA level 1 end states (i.e. Plant damage states) can be used as initiating point of severe accident, with small review for the plant damage states which was excluded from PSA due to very low frequency.
- With use of PRA containment event trees it is possible to identify the severe accident plant end states which corresponds PRA level 2 release categories and after that determine the severe accident scenarios by tracking Containment Event Tree branches with failures of severe accident management measures and operator actions.
- The ultimate severe accident list is to include the accident scenarios which corresponds to each level of plant release (plant severity level) in the generic event trees.

### 2.4.9. Summary of IRSN's experience on the development and application of Level 2 probabilistic safety assessment, France

IRSN has initiated the development of Level 2 PSA since the mid-90s and today uses a set of tools and methodologies able to build some 'best estimate' accident progression event tree, to quantify uncertainties, and to compute the radioactive releases and the radiological consequences for a large set of accident scenarios. These tools include for example the integral code ASTEC, specific codes like MC3D and Cast3M, the probabilistic code KANT or fast-running codes for radioactive release calculations and their consequences (MER, MERCOR).

The lessons learned from the Level 2 PSA developments have been used for the periodic safety review performed for French NPPs in France since 2008 and the interest of plant modifications dealing with severe accident have been confirmed by the knowledge acquired with these developments.

Concerning methodologies for HRA for actions specifically dedicated to accident management guideline strategies, IRSN has built and applied the HORAAM method. The HORAAM method has been developed by IRSN in order to predict the human error probabilities (HEPs) of the actions contained in the severe accident management guidelines (SAMGs) of the French 900 MW(e) PWRs. HORAAM is a decision tree model, where the HEPs are evaluated through a limited number of influencing factors that define the context for the action. HORAAM has gone through a number of steps in its development. The first one was the observation of crisis centre exercises, in order to identify the main IFs which, affect human and organizational reliability in a situation of core damage. Then crisis experts were asked to discuss the relevance of the selected influencing factors and to quantify their relative weight. It is planned to improve this HRA model for external hazards accident scenarios in the near future.

IRSN also presented some on-going research to support severe accident strategies, for example concerning ex-vessel corium stabilization, iodine filtration in case of filtered containment venting system opening or ASTEC capabilities improvement.

The low power and shutdown states are modelled with the same approach that full power state. In particular, ASTEC is used to calculate these transients.

Thanks to IRSN developments on the Level 1 and Level 2 PSAs interface methodology, the introduction of internal hazards and some external hazards is as simple as when dealing with internal initiating events (earthquake is an exception). Moreover, the HORAAM method is flexible and allows adaptation of human actions success probability depending on the hazard's context.

#### 2.4.10. ASME/ANS Level 2 standard on probabilistic risk assessment[8]: Status and preview of the standard, United States of America

The Level 2 PRA standard is a joint product of ANS and ASME and is managed under the Joint Committee on Nuclear Risk Management. This standard is a revision to the Ref. [38] trial use version and addresses lessons learned from its trial use. The revised standard provides criteria and acceptable methods for the evaluation of containment performance and radiological releases to the environment which result from postulated accidents that cause fuel damage in current generation and advanced light water reactors operating in the United States. The scope of this standard is to define requirements for analysing the progression of severe accidents, starting from core damage initiation to radionuclide release into the environment. It covers phenomena occurring within the reactor vessel, containment structure, and neighbouring structures that could contribute to radiological releases. The analysis involves assessing postulated accident sequences using probabilistic logic structures, such as containment event trees, to determine radionuclide release characteristics (e.g. magnitude and timing) through various pathways. The Level 2 standard is intended to be applicable to reactor accidents initiated from all modes of reactor operation (at-power, shutdown, and transition states) and during a full spectrum of accident sequences initiated by internal events and/or external hazards. The standard does not address accidents originating in the SFP. A genericized version of this standard has been developed and included within the PRA standard for non-LWRs.

The Level 2 standard builds upon Refs [39, 40] and enables a refined assessment of LERF along with requirements for the assessments of other plant release states including those with small radiological releases and large late radiological releases. The results of the Level 2 analyses may be used to assess containment end states and/or may be used as input to Level 3 consequence analyses.

Consistent with the ASME/ANS suite of PRA standards, the Level 2 standard provides a seamless interface and common structure to the Level 1 standard. This common structure includes common definitions, a parallel format for identifying high level and supporting

[8] The term of Probabilistic Risk Assessment (PRA) is equivalent to Probabilistic Safety Assessment (PSA)

requirements and parallel document sections requirements for expert judgment, peer review, and PRA maintenance and update.

While the focus of the standard is operating plants in the United States, the technical requirements need to be equally applicable to LWR designs, worldwide. The detailed structure and organization of the standard were presented.

The standard revision process is ongoing. Resolution of trial and pilot use application comments is still in progress and structural changes to the standard have been made to ensure consistency with the Level 1 standard.

### 2.4.11. Performing thermohydraulic calculations within Level 2 probabilistic safety assessment for shutdown modes, Russian Federation

Shutdown modes can be an additional source of potential danger due to transient and specific technical procedures. For this reason, additional attention needs to be paid to them during safety analysis.

According to Ref. [14], safety analysis needs to be performed for all plant operating states, taking into account all the locations of nuclear materials, radioactive substances and radioactive waste at the NPP, in which abnormal conditions of the NPP may occur. The following values are set as a target for NPP safety:

- The total probability of severe accidents for each NPP unit on an interval of one year may not exceed $1 \times 10^{-5}$.
- The total probability of large release for each NPP unit on an interval of one year may not exceed $1 \times 10^{-7}$: at the same time, a large release is determined taking into account the permissible doses for the population.

The abstract presented the experience of the PSA performed for the 2nd unit of the Leningrad 2 NPP.

According to Russian Federation standards and requirements for granting the licence to start commercial operation, it is necessary to perform a full scope Level 1 and Level 2 PSA, which includes:

- Assessment of internal initiating events;
- Initiating events caused by on-site hazards (e.g. fires and floods);
- Initiating events caused by natural and human made external hazards.

A full scope Level 1 and Level 2 PSA corresponding to the as-built stage of Unit 2 was presented to the Russian regulatory body in early 2020 and, after successfully examination, the 2nd unit of Leningrad II NPP was put into commercial operation.

Level 1 PSA results showed a significant contribution of shutdown modes to the frequency of nuclear fuel damage. Among them, there are three important scenarios for planned cooldown from a safety point of view:

1. Simultaneous failure of the residual heat removal system and reactor make-up system with a sealed primary circuit.
2. Simultaneous failure of the residual heat removal system and reactor make-up system with an unsealed primary circuit.

3. Station blackout with an opened reactor.

The first scenario is dangerous because it sets the maximum mass and energy release that can be obtained into the containment for the shutdown modes, the maximum corium release into the core catcher, and the need to reduce the pressure of the primary circuit. The second and third scenarios are dangerous due to the delayed operation of the isolation valves and the maximum radioactive release inside the containment.

These scenarios have a similar chronology to events that have occurred: there is a failure of the residual heat removal system due to a mechanical reason or due to station blackout, including failures of supporting systems leading to a dependent failure of the residual heat removal system. Due to decay heat in the core, the coolant heats up and boils off, the core heats up and fuel element damage starts. The next step is core melting.

The thermohydraulic calculations for the scenario with a sealed primary circuit (scenario 1) showed that the operation of two safety valves of the residual heat removal system is not enough to reduce the pressure in the primary circuit to prevent core meltdown ejection at high pressure. Thus, according to severe accident management guide when reaching a temperature 400°C above the core, the personnel need to open two out of three pressurizer safety valves and one out of two emergency gas removal system valves (to the containment or to the relief tank) to reduce the primary circuit pressure to 1 MPa.

In the 2nd and 3rd scenarios, a reactor is opened with a low coolant level (the level is 550 mm above the axis of the inlet 'cold' branch pipe). The radioactive substances released from the damaged fuel elements easily enter to inner containment volume and only a small part is retained in the primary circuit. Therefore, it is important to evaluate the success of the isolation functions of the containment ventilation systems on the signal of overpressure up to 0.105 MPa due to boiling off.

Before performing thermohydraulic calculations, it was unknown if the pressure inside the containment could increase to the set point level, because the ventilation system provides a high rate of cooling and pressure reduction (scenario 2). In the shutdown modes, as opposed to power operation modes, the high-capacity ventilation system is in operation.

In case of the 3rd scenario, with a station blackout, the ventilation system fails and a containment bypass with a hydraulic diameter of about 1000 mm is formed.

The thermohydraulic calculations showed that the containment is isolated before the beginning of the core top draining. Thus, according to calculations, the selected capacity of the batteries for powering the isolation valves is acceptable.

An analysis of the accident sequences with the reactor core destruction and the corium release outside the reactor vessel was performed using a set of special programs and codes: SOKRAT/V3, KUPOL-M, LIMITS-V, FIRECON 2.0, KIN and DOZA 3.0.

Further results are obtained in three calculation groups:

- The obtained parameters of the mixture (concentrations of oxygen, hydrogen and steam, and the pressure and temperature) are used to model and determine the combustion modes of the hydrogen-air-steam inside the containment.
- The sources of masses and energies of water, water steam and non-condensed gases obtained are used for the analysis of containment continuous loading.

- The obtained fission products sources together with other previously obtained parameters are used to calculate the transfer of fission products in the containment rooms and beyond.

As a result, the obtained values of releases and scattering parameters are used for release categorization and calculation of doses received by the population.

In the PSA model, the results were taken into account when developing the containment event trees and system fault trees.

The need to reduce the primary circuit pressure by the personnel using pressurizer safety valves and the emergency gas removal system is taken into account in the event tree, which is modelling a severe accident with the loss of the residual heat removal system in the shutdown modes with a sealed primary circuit. Also in the containment event tree, we are taking into consideration the following results of thermohydraulic calculations: the pressure inside the containment, the combustion or detonation of hydrogen-air-steam mixtures, the core catcher, etc.

The need to close the containment isolation valves by an automatic signal when the pressure increases, is taken into account in the event tree, which is modelling a severe accident with the loss of the residual heat removal system in the shutdown mode with an unsealed primary circuit.

Analysis of the accident sequences demonstrates that the personnel need to open two out of three pressurizer safety valves and one out of two emergency gas removal system valves in the case of the loss of residual heat removal system (regardless of the operation of the residual heat removal safety valves) to prevent core meltdown ejection at high pressure.

Calculation results demonstrate that in the case of an accident during shutdown mode with an opened reactor, 0.105 MPa pressure in the containment is reached at the initial stage due to evaporation of the coolant and the isolation valves being successfully closed automatically before the batteries discharge and before the reactor core begins to be exposed. Thus, the protective measures provided in the design exclude the direct release of radionuclides into the environment and prevent large scale contamination of vast areas.

At the next stage of Level 2 PSA development, more detailed modelling of processes inside the containment is planned for the case of a severe accident in the in-containment fuel pool. Currently, the conservative assumption is that all severe accidents in the fuel pool lead to a large release.

### 2.4.12. Use of severe accident codes to refine the large early release frequency to support realism in estimates, United States of America

The current emphasis on efficient operation of the US NPPs along with the approval of risk informed tools has opened opportunities to optimize aspects of operation. Performing these studies requires continuous improvement of plant PRA models. The current emphasis is centred on internal event and internal fire core damage frequency refinement and inclusion of diverse and flexible mitigation capabilities (FLEX). The large early release frequency modelling in many cases is still based on a simplified or bounding assessment using generic implementations of large early release frequency models without a specific source term or other characterizations. For some applications, this has been demonstrated to lead to biased results and the potential for not meeting regulatory targets.

This presentation presented an approach for refinement of large early release frequency model outcomes based on the decomposition of large early release frequency into the two elements: early and large.

The approach for refinement of large early release frequency based on the definition in Ref. [41] and radionuclide source term timing at a high level involves the decomposition of large early release frequency into the two elements: early and large. Calculated accident scenarios are removed from more generic large early release frequency classifications based on failing to meet both criteria.

The definition of 'early' is refined based on the plant-specific operational guidance combined with an assessment of accident progression timing. The refined assessment considers the time when a site general emergency is declared to initiate evacuation that is sequence specific and then applies the time elapsed between the initiation of evacuation of the general population surrounding the site until the area is evacuated (95% of the surrounding population). The basic time equation is written as follows:

$$T(EARLY) = T(0)+Tac(GE)+T(IMP)+T(EVAC) \quad (1)$$

Where:

- T(0) represents the initiating event;
- Tac(GE) is the time to a general emergency defined on an accident sequence (ac) basis applying the core damage model results to the current emergency action guidelines (EALs);
- T(IMP) is the time required to implement the evacuation order;
- T(EVAC) is the time required to evacuate the emergency planning zone.

Figure 11 illustrates the implementation of the timing elements to arrive at the breakpoint between early and late. The point where evacuation is completed is the realistic break point defining success at precluding prompt fatalities.

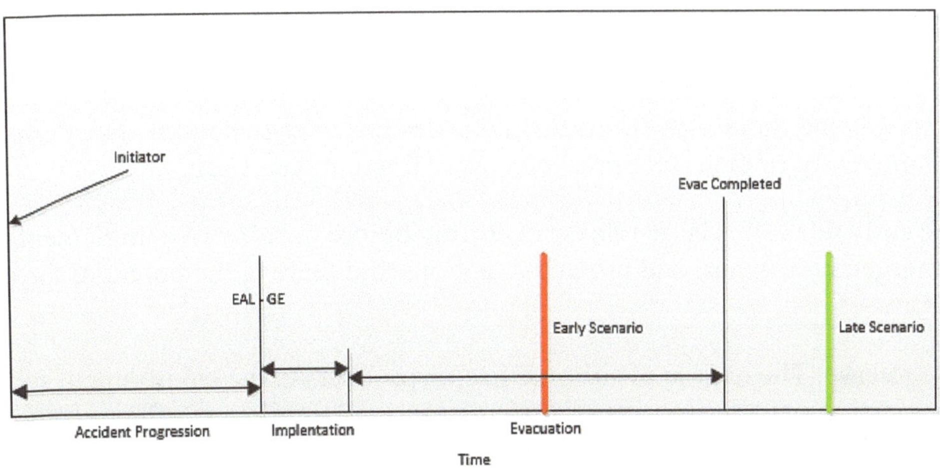


*FIG. 11. Accident response timing study (courtesy of R. Summit, Engineering Planning and Management, United States of America).*

The times can either be based on a scenario or conservatively estimated as the shortest time element. The implementation and evacuation windows will not vary substantially for a specific hazard. However, some hazards (tornado, seismic, etc.) can impact on the ability of the public authority to evacuate the public and have to be included.

The definition of 'large' is refined based on a review and remapping of accident progression and radionuclide constitution and release timing. A refined definition for large is made considering the composition of the release regarding health effects and the retention or holdup of radionuclides leading to reduction or delays in the release of radionuclides to the environment. A review of previous studies reveals that the caesium and iodine release fraction can be used for off-site consequences. That is, an approximate relationship exists between the fraction of caesium and iodine released and the population dose.

A reclassification scheme is based on having enough radionuclides to generate early health effects and large is when these radionuclides exceed 3% of the core inventory. The scenarios shown to not meet this criterion can be screened from the large early release frequency. Credit is also taken for systematic scrubbing and deposition and holdup within the containment.

The application of this approach provides an ability to move the classification of large early release frequency from a more generic to a more site-specific assessment leading to reclassification of scenario frequency from large early release frequency to non-large early release frequency and an overall reduction in large early release frequency based on the inclusion of more realism in the analysis.

The accident scenarios are tested against each element (large, early) based on severe accident analysis and plant specific accident management processes and binned according to the results. Large early release frequency scenarios have to meet both criteria.

### 2.4.13. Summary of Session IV

It can be concluded from the national presentations and discussions that Member States have made efforts to consider hazard and their combinations, and modelling severe accident management guidelines in Level 2 PSA. During the discussion on severe accident management guidelines modelling in Level 2 PSA, the importance of considering the possibility of worsening the accident conditions while implementing severe accident management guidelines (i.e. side effects of the severe accident management guidelines actions) was highlighted.

The ASME/ANS Level 2 status and preview of the standard was presented. During the discussions, it was mentioned that the standard is not technology neutral. The definitions large release and large early release were presented as defined in Ref. [42]:

> "Large early release: A large release occurring before the effective implementation of off-site emergency response and protective actions and there is the potential for early health effects.
>
> "Large release: The release of airborne fission products to the environment such that there are significant off-site impacts. Large release and significant off-site impacts may be defined in terms of quantities of fission products released to the environment, status of fission product barriers and scrubbing, or dose levels at specific distances from the release, depending on the specific analysis objectives and regulatory requirements,"

It was also underlined that the standard focused on core damage and the SFP was outside of the scope when the activities were initiated.

The presentation of EMP, USA provided an approach for the LERF estimate, based on the US NRC definition of the Prompt Fatality Quantitative Health Objective.

During the discussion, it was revealed that in the study performed by the Korean Atomic Energy Research Institute, the non-permanent equipment was introduced in the Level 1 and Level 2 PSA models without modelling the severe accident management guidelines procedure. In future, the efforts will be spent to model severe accident management guidelines in the PSA. It was emphasized that during the integration of the severe accident management guidelines in the PSA special attention needs to be paid to the fact that in the implementation of the severe accident management guidelines the accident conditions might worsen, and it needs to be reflected in the model.

Also discussed was the need to consider the water level drop in the site caused by a tsunami, which can cause rupture of the cooling system and trip pumps, and cause missiles from the debris brought to the site by the tsunami wave.

The approach used by IRSN regarding the analyses of the Level 1 and Level 2 PSA operator errors dependencies was discussed. It was emphasized that the PANAME method is used for Level 1 PSA at IRSN. PANAME stands for "Plan d'Actions Nouvelles pour l'Amélioration du Modèle EPFH avec passage à l'Approche par Etats", which can be translated in English as "a HRA method improved to take into account the state-oriented procedures". PANAME is a 'THERP family' method. The first part of the modelling takes into account the crew in the main control room. The second part is dedicated to the recovery possibilities by additional staff. The safety engineer's mission is the monitoring of the most important actions with regard to safety. PANAME had been developed by IRSN on the basis of the HRA method FH6 (developed by Électricité de France (EDF) before MERMOS) for which EDF implemented a large amount of simulations from a full scope simulator to obtain data of personnel using state oriented procedures. The main characteristics of PANAME are:

- Utility data (from EDF);
- Time correlation curves for the diagnosis;
- Decision trees for selecting the context factors and the recovery probabilities;
- An additional tool for the analysis of procedure paths;
- A computerized worksheet is available to support the modelling.

In the severe accident management guidelines, two types of action are considered, immediate actions and differed actions; in case of the differed actions the national crisis team is required. For immediate action, the PANAME method is used, meanwhile for differed actions when the national crises team is required the HORAAN method is used. The following shaping factors are used to estimate human error probability: decision time, available information and measures, decision difficulty (e.g. venting of the containment), the difficulty for the operator in the control room, scenario difficulty. It was mentioned that available information and measures on the Level 1–Level 2 PSA interface are also considered during the implementation of HRA. The decision time is estimated based on the severe accident entry time and the time the initiating event occurred.

During the discussion, it was agreed that the early release definition can be different for the same site depending on the external hazard. This is conditioned by the fact that the time window for evacuation can differ based on external hazards (e.g. in case of seismic hazards, the roads might be affected and the emergency response service will be overloaded).

## 2.5. SESSION V: INDUSTRY PERSPECTIVE

### 2.5.1. Session V description

The last session contained practical experience on Level 2 PSA from nuclear industry organizations such as NPP operating organizations or groups. The presentations reflected different aspects, such as application of PSA methods in the risk decision making process, development of Level 2 PSA for the spent fuel pool, development of deterministic studies in support of Level 2 PSA for sodium cooled reactors and an overview of the current status of development of Level 2 PSA for different NPPs. The abstracts of presentations are presented in Section 2.5.2–2.5.9. The discussion summary is presented in Section 2.5.10.

### 2.5.2. Application of Level 2 probabilistic safety assessment to sodium cooled fast reactors, Japan

The International Forum for Generation IV reactor systems states in the basic safety approach that Generation IV designs are developed from the earliest stages in a manner that is guided by knowledge derived from PSA and other formal safety assessment methods. Recently, a risk informed, performance based approach [43, 44], developed in the United States has started to be applied to the safety design of non-LWR advanced reactors. Level 1 and Level 2 PSA methodologies have been developed and are going to be applied to the design process of advanced sodium cooled fast reactors (SFR) in Japan.

During its long history of worldwide SFR development, event progression and the consequences of hypothetical core disruptive accident have been studied. A lot of experimental data has been accumulated and analysis tools have been developed and applied to actual SFR reactor designs. Japan has been participating in such activities and has obtained knowledge. The basis of a PSA methodology, tools and database for SFRs has been established by conducting a full scope Level 1 and Level 2 PSA for the prototype SFR Monju, and the results were used for its safety review and formulation of accident management procedures.

In the conceptual design work of the Japan SFR as a Generation IV reactor, mitigation measures against core damage are introduced as level 4 defence in depth, for which a Level 2 PSA methodology is developed. In general, SFR event sequences leading to severe accidents can be categorized into two major groups; anticipated transient without scram (ATWS), and loss of heat removal system (LOHRS). Design measures against ATWS, which might end severe core damage, are intended to contain the degraded core inside the reactor vessel (IVR). Design measures against LOHRS are intended to maintain the liquid sodium surface level sufficient above the core and the heat transfer path open to the ultimate heat sink. In order to evaluate the efficiency of these design measures, the application of Level 1 and Level 2 PSA is useful.

For ATWS sequences, it is necessary to evaluate the efficiency of the core damage prevention capability of inherent reactivity feedback and passive reactor shutdown mechanisms. For this purpose, phenomenological analysis of various forms of reactivity feedback, such as Doppler effect, fuel thermal expansion effect and passive mechanisms, including uncertainty analysis are required. In the subsequent sequences, in the case of core damage progression (Figure 12a), the extent of core damage and the time scale of core damage progression vary depending on the postulated event spectrum. In general, SFR core damage sequences inside the reactor vessel are analysed as a series of accident phases, i.e. primary phase, transition phase, relocation phase and post-accident cooling phase. The failure probability of IVR is evaluated as the result of the individual phase analysis.

For LOHRS sequences, a sufficient number of design measures with diversity needs to be provided so that core exposure and complete loss of decay heat removal can be prevented even under severe plant conditions. The capability and reliability of these design measures are analysed in relation to the resultant core temperature increase and cladding damage fraction (Fig. 12b).

The resultant radioactive release from the reactor core and its transport to the environment are analysed and summarized in the event trees, which include the in-vessel part and containment part. This can be used for the evaluation of mitigation effects on a radioactive release.

Utilizing all the above-mentioned results, in addition to deterministic consideration, the design of Generation IV SFRs can proceed from conceptual design to basic design and then detailed design. In parallel, R&D work will be done to extend experimental data to reduce the uncertainty associated with highly important phenomena.

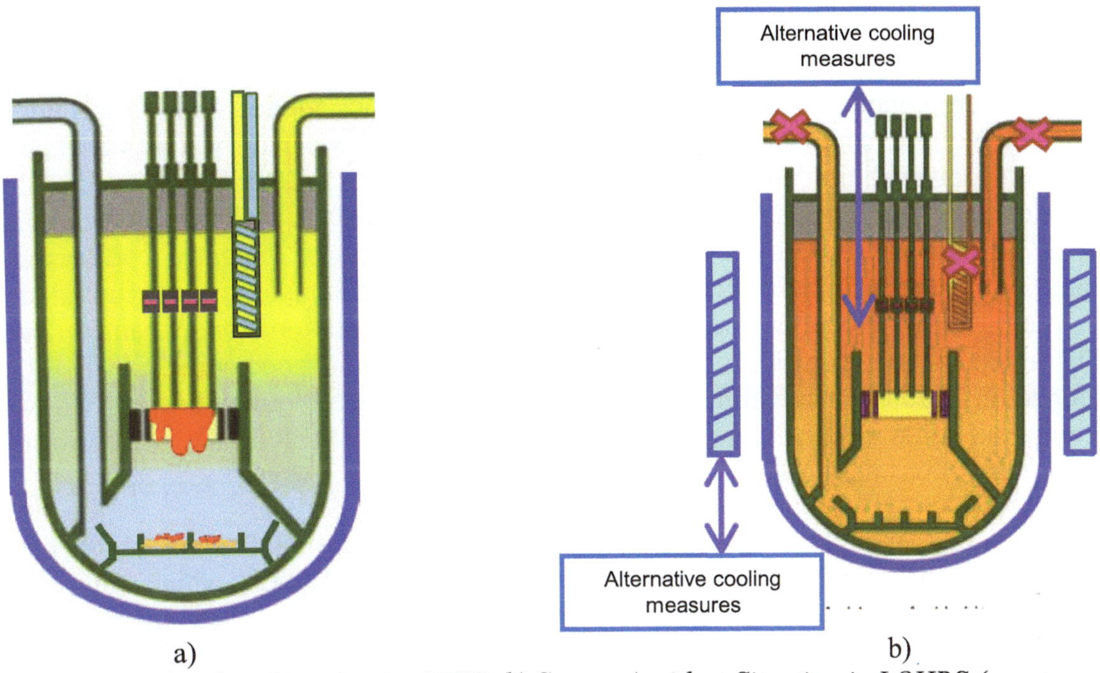


*FIG. 12. a) Severe Accident Situation in ATWS; b) Severe Accident Situation in LOHRS (courtesy of S. Kubo, Japan Atomic Energy Agency, Japan).*

### 2.5.3. Development of severe accident simulation methodologies for sodium cooled fast reactors, Japan

In general, SFR event sequences leading to severe accidents can be categorized into two major groups; ATWS and LOHRS. Design measures against ATWS, which might lead to severe core damage, are to contain the degraded core inside the reactor vessel, which is called the IVR. Design measures against LOHRS are intended to maintain the liquid sodium surface level sufficiently above the core and the heat transfer path open to the ultimate heat sink. Severe accident assessment against the ATWS is of prime importance from the viewpoint of safety characteristics of SFRs, which have a potential re-criticality leading to significant mechanical energy release. JAEA has been developing severe accident simulation methodologies for SFRs, which can be applied to Level 2 PSA.

For ATWS, in-vessel accident sequences are evaluated in four phases: initiating, transition, material relocation and heat removal phase. JAEA has developed the SAS4A computer code for the initiating phase, the SIMMER code for the transition and material relocation phases, and the Super COPD with a debris bed module for the heat removal phase. The CONTAIN-LMR code has been developed for ex-vessel accident sequences. This presentation described these severe accident simulation methodologies, as shown in Fig. 13.

In SAS4A, a channel consisting of one representative fuel pin (fuel pellet and cladding), coolant path and wrapper tube, is used to model fuel assemblies. In steady state calculations, fuel restructuring with irradiation and fission gas accumulation in fuel pellets during normal operation are analysed. In transient calculations, this code can calculate important phenomena in the initiating phase such as sodium boiling, fuel pin failure, axial expansion of the fuel pin failure region, fuel motion and fuel coolant interaction after fuel failure. Based on material and temperature distributions obtained by the transient analysis and reactivity profile data specified by SAS4A input data, the reactivity of sodium void, fuel axial expansion, fuel motion (fuel dispersal), structure and Doppler are calculated. Using the point kinetics equation, the reactor power transient with time is calculated.

The SIMMER-III operates as a two-dimensional, multi-component, multi-phase, Eulerian fluid dynamics code, coupled with fuel pin model and neutronics model. The code underwent a comprehensive and systematic assessment in two phases as part of the verification and validation. During the first phase, fundamental assessments focused on single- and multi-phase flow benchmark problems, small scale experiments using reactor and simulant materials, and physical problems with known solutions. In the second phase, integral code assessments explored complex multiphase scenarios relevant to accident analysis. These included boiling pool dynamics, fuel relocation and freezing, fuel–coolant interaction, core expansion dynamics, and disrupted core neutronics. Additionally, a three-dimensional code called SIMMER-IV was developed in parallel, maintaining the same physical model framework as SIMMER-III. Reactor application studies were also conducted using SIMMER-III/IV, demonstrating the reliability and robustness of the code.

The debris bed module is implemented in a one-dimensional plant dynamics code, Super-COPD. The scope of this module pertains to the debris bed, which comprises solid particles (fuel and steel) and sodium coolant (liquid/vapour). The module calculates temperature and saturation distributions within the debris bed using fundamental equations applicable to both subcooled and boiling regions. However, it does not model the movement or melting of the debris bed.

The CONTAIN-LMR code calculates the behaviour in an SFR containment vessel, with coupling together the various phenomena of thermohydraulics, aerosol particle behaviour, and radionuclide transfer behaviour. For example, the code has the sodium-coolant-specific models of sodium evaporation, condensation, combustion, aerosols behaviour, and sodium–concrete reaction, as well as the general models of heat transfer, gas flow in a multi-cells system, hydrogen burning and so on which can be used independently on the kind of the coolant. The sodium related models in CONTAIN-LMR have already been individually validated with specific experiments for those such as sodium fire, aerosol behaviour, and sodium–concrete reaction.

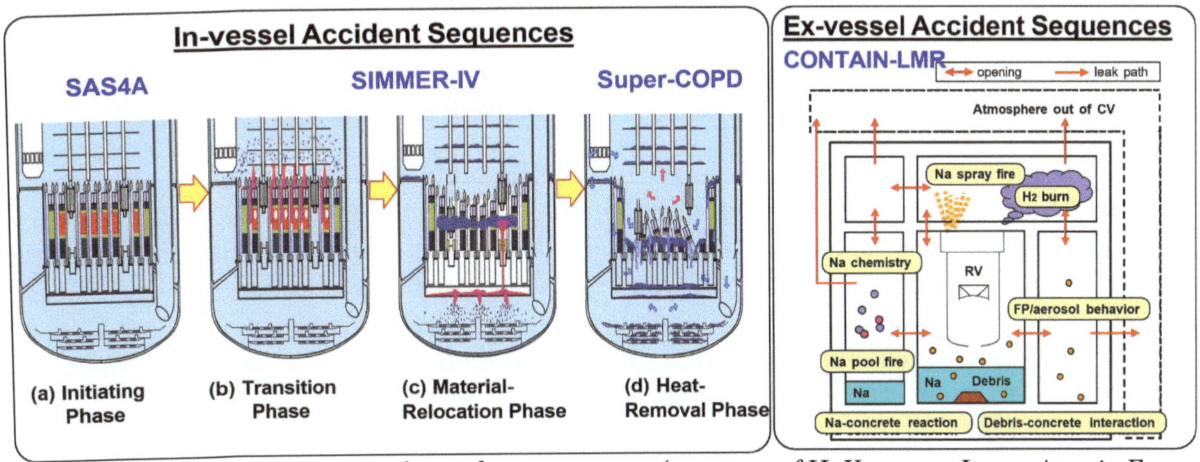


*FIG. 13. In-vessel and ex-vessel accident sequences (courtesy of H. Yamano, Japan Atomic Energy Agency, Japan).*

### 2.5.4. Overview of Level 2 probabilistic safety assessment for the Cernavoda nuclear power plant, Romania

Cernavoda NPP is owned and operated (licence holder) by the National Company Nuclearelectrica (Societatea Nationala Nuclearelectrica, further referred to as SNN). At this time, the site has two operating plants, Cernavoda Unit 1 and Unit 2.

In addition to the deterministic approach, probabilistic analyses have been performed for both units in operations at Cernavoda NPP. Such an assessment of the adequacy of plant-specific risk provides both a measure of potential accident risks to the public and insights into the adequacy of plant design and operation.

The scope of the existing PSA covers Level 1 and Level 2 for internal events and hazards (fire and flood) for all operating states. A PSA based seismic margin assessment has also been performed.

A specific strength of the existing PSA resides in the close collaboration between a dedicated plant personnel PSA team, plant designer and internationally recognized experts who provided support as part of the team or via various IPSRT and other support missions organized under technical cooperation programmes coordinated by the IAEA. Under these circumstances, the Cernavoda PSA team efforts for continuous improvement has been supported by both plant management and international organizations. Use of the most realistic assumptions and consideration of the latest methodology development ensure the quality of the existing studies and day by day use for risk informed decision making in the plant processes such as work planning, operating risk reduction, design changes and personnel training.

The methodology for conducting a Level 2 PSA is based on internationally recognized practices and consistent with the requirements and recommendations in IAEA safety standards. In addition, this methodology document draws extensively on the experience from the successful performance of a Level 2 PSA performed by AECL/CANDU Energy (CEI) for other CANDU 6 NPPs.

The Level 2 PSA was performed in accordance with the Level 2 PSA methodology prepared by CEI. The objectives of the study were inspired by the objectives presented in para. 2.5 of SSG-4 [2], which states:

(a) “To gain insights into the progression of severe accidents and the performance of the containment.

(b) “To identify plant specific challenges and vulnerabilities of the containment to severe accidents.

(c) “To provide an input into the resolution of specific regulatory concerns.

(d) “To provide an input into determining compliance with the probabilistic safety goals, or with probabilistic safety criteria if these have been set. Typically, such probabilistic safety goals or criteria relate to large release frequencies and large early release frequencies.

(e) “To identify major containment failure modes and their frequencies and to estimate the associated frequencies and magnitudes of radionuclide releases.

(f) “To provide an input into the development of strategies for off-site emergency planning.

(g) “To evaluate the impacts of various uncertainties, including uncertainties in assumptions relating to phenomena, systems and modelling.

(h) “To provide an input into the development of plant specific accident management guidance and strategies.

(i) “To provide an input into determining plant specific options for risk reduction.

(j) “To provide an input into the prioritization of research activities for the minimization of risk significant uncertainties.

(k) “To provide an input into Level 3 PSA consistent with the PSA objectives.

(l) “To provide an input into the environmental assessment of the plant.”

With the addition of specific objectives referencing Romanian regulatory documents, such as; to provide an input into determining compliance with the probabilistic safety goals, or with probabilistic safety criteria, that relate to large release frequencies and large early release frequencies as defined in the Romanian regulatory documents.

The Level 2 PSA focuses on assessing potential accident sequences that may occur after core damage. It examines phenomena that could challenge containment integrity and potentially result in the release of radioactive material into the environment.

The two most important criteria used to characterize the radiological release are:

(a) The estimated magnitude of the total release;

(b) The timing of the first significant release of radionuclides.

The predicted source term associated with each release category, including both the timing and magnitude of the release, is determined from the accident progression analysis using MAAP4-CANDU.

For Level 2 PSA, the acceptance criterion is defined in Level 2 PSA methodology Cernavoda 2, and it is based on the recommendations from Ref. [45].

According to this document, the objective for large off-site releases requiring short term off-site response is $1 \times 10^{-5}$ events per reactor-year for existing plants.

Details of the methodology and results of the Level 2 PSA for Cernavoda NPP were detailed in the extended presentation.

### 2.5.5. Contextual interaction between Level 1 and Level 2 probabilistic safety assessment for risk informed optimization of Kozloduy nuclear power plant project and technical specifications, Bulgaria

PSA serves as an integrated tool for understanding and communicating risk profiles, for determining priorities, effectiveness and correctness of decision-making and safety measures.

Taking into account statistics after the accident at the Fukushima Daiichi NPP, it could be said that the CDF, LERF and safety measures against severe accidents are underestimated. This underestimation could be explained by missing information, insufficient analyses for external and common cause events, multi-unit dependencies, inadequate training, design flaws, inadequate safety measure and upgrades, lack of safety culture. The common reason for all of them is that the previous analyses are fully or partially outside the scenario context and its dynamics. Hence, deterministic safety assessment (DSA) and PSA together have to prepare better comparable tools, realistic models and usable risk informed applications, simulations and experiments in order to improve data use and methodologies, to reduce the uncertainty of static results and give a dynamic contextual interpretation for them. Static and non-contextual safety analyses are used entirely for the PSA and partly for DSA methodology development and its applications. Therefore, an evaluation procedure for dynamic context qualification and quantification needs to be applied to complement and compensate the weaknesses in the interaction between all analyses of safety and especially between Level 1 PSA and Level 2 PSA. Without such interaction between them, as well as with DSA, many of the possible PSA applications will not be effective, the uncertainty of the results will not be reduced, and the trust in the PSA will not grow.

The presentation provided the prospective areas and six examples for on-going Kozloduy NPP applications for the contextual interaction between PSA Level 1 and Level 2.

The prospective areas for using PSA in the Kozloduy NPP operation, maintenance and repair are:

(a) Integrated risk informed decision making process.

(b) Risk informed technical specifications (RITS) by PSA Levels 1 and 2:

  (i) Justification of acceptance rules for RITS changes;

  (ii) Areas of compliance for justification the configurations changes of safety systems.

(c) Justification of the risk informed AOT (i.e. allowed outage times) changes for risk informed operation, maintenance and repair.

(d) Risk informed non-destructive testing programmes and in-service inspection.

(e) Risk informed equipment testing programmes.

(f) Analysis of operational events and safety measure improvement.

(g) Crew performance evaluation.

(h) Staff training programmes.

(i) Symptom-based emergency operating procedures (EOPs).

(j) Severe accident management guidelines (SAMG).

(k) Changes in the design of the SSCs.

(l) Deterministic–probabilistic safety classification of equipment (SSC).

(m) Changes in operating limits (SL, OL) and conditions.

(n) Risk and project balance demonstration.

(o) Unit project evaluation within the periodic safety review (PSR).

The on-going Kozloduy NPP applications for the contextual interaction between Level 1 and 2 PSA are exemplified by:

Example 1 – Deterministic–probabilistic safety classification of structures, systems and components.

Example 2 – Thermohydraulic, context and joint criteria for safety measure optimization of VVER-1000 station blackout scenario.

This application shows the opportunities for improvement, enhancement, update and optimization of the severe accident strategies, safety measures, simulation codes capabilities, EOPs and SAMGs using insights from contextual interaction between Level 1 and Level 2 PSA models. A procedure for dynamic context quantification and determination of alternatives, coordination and monitoring of human performance and decision making is proposed and applied in order to:

- Minimize, monitor and control the potential probability and/or impact of unfortunate events;
- Maximize the realization of opportunities or alternatives.

The procedure is exemplified for eight station blackout scenarios of the NPP with VVER-1000 based on different codes and models: ATHLET/Mod1.2/2.2, RELAP5/MOD3.2, MELCOR 1.8.5, and ASTEC V1.3/V2. The study of the overall context of these scenarios makes it possible to take into account not only the thermohydraulic characteristics, but also the complexity of the contexts arising in mastering the situation. The results of the context study complement the thermohydraulic analyses and provide additional opportunities for evaluation of alternative strategies.

Example 3 – Context-based digraph decision making model during the emergency situation on the NPP site.

The third application offers a model for assessing the organizational structure for emergency decision making based on the context. The data are necessary for correct evaluation of considerations and context symptoms related to modelling of any human and crew performance (supervisor, manager, head, group, organization, etc.) in Level 1 and 2 PSA.

Example 4 – The Kozloduy NPP proposal for the risk informed operation, maintenance and repair acceptance rules.

The experience of Kozloduy NPP operation has shown that both planned and unplanned repair and maintenance play a very important role in the process of safety management. One of the beneficial ways of enhancing Kozloduy NPP safety is to improve and optimize the activities and requirements in the technical specification (TS) by considering risk informed decisions in addition to deterministic evaluations and expert opinion. Furthermore, additional analyses are necessary to determine which POS has to be chosen for repair and maintenance of certain equipment.

Kozloduy NPP has developed a proposal for the evaluation of changes of the risk informed AOTs and non-risk informed requirements. The proposal uses two kinds of curve that have to be matched by Refs [41, 46] acceptance rules for both conditional core damage probability and conditional large early release probability:

(1) Pareto frontier of undominated choices, determined by the physics and economics of the situation, unrelated to subjective preferences of the decision maker;

(2) Decision maker's compromise (utility trade off) curves, which are independent of particular choices available.

Example 5 – Areas of compliance for the Kozloduy NPP risk informed (RI) TSs.

By the use of the risk monitoring software tool (RS Risk Watcher) based on the Level 1 PSA model and with the support of the Level 2 PSA model, the contributions to the respective risks (CDF, LERF) were calculated from the planned or unplanned repair and maintenance of many safety related system configurations, according to the required AOT by TS (3 days). The goal was to find opportunities for the risk informed optimization of repair and maintenance activities, as well as to change the RITS.

Level 2 PSA has been applied to evaluate the contribution of using the permanent equipment to cope with allowed outages. In the future, one of the objectives of Level 2 PSA will be to assess the contribution of the portable equipment to the management of severe accident situations and its role in taking compensatory measures during repairs and maintenance. Such an assessment can also serve to improve education and training of the staff working with this equipment.

Options for the limited AOT of the unplanned repair are possible in four of the plant operating states according to the non-RITS requirements for AOT $\leq$ 3 days of the $SSC_i$ of the Kozloduy NPP: a) POS0 ('full power'); b) POS2+ ('hot shutdown'); c) POS6+ ('cold shutdown'); d) POS10+ ('no fuel in the reactor'):

An option for the unlimited AOT (i.e. more than 3 days) of the planned repair could be allowed only when the increase in CDF for and specific POS (e.g. no fuel in the reactor) complies with the non-risk-informed TS requirement. This is the safest choice but if some groups of SSCs (channels with power and water supply) aggregate a high cumulative risk (CDF/LERF), then they have to be ungrouped and repaired separately as for the unplanned repair above.

As a result of the optimization of AOT, the following four RI areas of compliance for justification and change of the non-RI TS requirements were identified:

Area 1: All AOTs for non-risk-informed repairs in this area need to be changed as risk informed, and the preferred order to perform the repair (unplanned and planned) is: $\Delta CDF_{POS0}$, $\Delta CDF_{POS10}$, $\Delta CDF_{POS2+}$, $\Delta CDF_{POS6+}$, $\Delta CDF_{POS10+}$;

Area 2: Non-RI AOTs in TS can be changed to RI for feasible repairs (unplanned and planned) and the preferred order to perform them is: $\Delta CDF_{POS0}$, $\Delta CDF_{POS10}$, $\Delta CDF_{POS2+}$, $\Delta CDF_{POS6+}$, $\Delta CDF_{POS10+}$;

Area 3: Risk informed repairs are feasible with additional corrective and compensatory measures, the preferred order to perform them is: $\Delta CDF_{POS10}$, $\Delta CDF_{POS0}$, $\Delta CDF_{POS2+}$ $\Delta CDF_{POS6+}$, $\Delta CDF_{POS10+}$;

Area 4: Risk informed repairs are not feasible without ungrouping the SSC and repairing them separately with design, corrective and compensatory measures in POS10 ($\Delta CDF_{POS10}$).

Example 6 – AOT calculation, selection and changing in TS by integrated risk informed decision making process.

Integrated risk informed decision making requires risk monitoring and reporting all contexts related to hazards, alarms, circumstances and violations. The contextual interaction between risk monitoring and Level 1 and 2 PSA models is an important feature that needs to be taken into account in calculating the actual and hypothetical risk curves of planned and unplanned repair and maintenance activities, corrective and compensatory measures. To be able to conduct an extended risk analysis and make balanced risk informed decisions, we need the PSA Level 1 and 2 multi-unit models of the NPP, which need to include the dynamic context of as many facilities, hazards and contributors on the NPP site as possible:

- Shared equipment, conditions and organization;
- Symptom-based SB EOP and SAMG;
- Common cause failures, events and hazards;
- Human failure events during the accident scenario progression.

### 2.5.6. Électricité de France Level 2 probabilistic safety assessment overview for operating plants, France

EDF's PSA scope has expanded considerably in recent years, in line with French regulatory requirements. Baseline Level 1 and Level 2 PSAs were only developed for internal events. In the framework of the fourth PSR involving the 900 MW(e) series (32 PWR units), Level 1 PSAs have been developed for internal hazards (fire, flooding, explosion) and external hazards (earthquakes, flooding) while Level 2 PSAs have been developed for fire, internal flooding and earthquakes events. Both Level 1 and Level 2 PSAs cover the at-power state, low power and shutdown modes, and consider fuel in the reactor building and in the spent fuel pool. The scope might be further expanded in the context of the next PSR to come (involving the 1300 MW(e) series, 20 PWR units) by integrating new external hazards and hazard combinations according to a screening process.

Level 2 PSA objectives and risk metrics definition are mainly based on the French regulatory context, EDF's objectives for the periodic safety review, and may consider para-regulatory context such as IAEA Safety Guides. It has to be noted that EDF Level 2 PSA does not aim to prepare any Level 3 PSA, which is not requested by the French regulatory body. Moreover, the French regulatory body does not define quantitative safety goals for Level 2 PSA. Instead, it requests continuous safety improvements during the plant operation lifetime.

For operating plants Level 2 PSA, release categories are defined on a 'symptom based' approach according to the degree of core damage, the containment status, the mitigating systems status. Simple radiological release bin categories are used for Level 2 PSA results presentation, representing an efficient way to ensure compliance with Level 2 PSA objectives and to capture insights. Since underground pollution is considered as a high stake in France, it has to be noted that each Level 2 PSA sequence has two different consequences: one for atmospheric releases and one for underground releases (basemat status).

An integrated PSA model approach has been used since 2008 using RiskSpectrum™ software. Some aspects of Level 2 PSA development process are specifically addressed in the presentation.

As regards to the PSA Level 1 and Level 2 interface, general principles for PDScet characterization were discussed. The difficulty lies in finding the trade-off between a precise characterization of accidental scenarios and the need to have a rather simple Level 2 PSA to be easily used and maintained. It has to be noted that PDS are also defined for some Level 1 PSA success branches that would be of interest for abnormal radiological release assessment (e.g. containment bypass sequences without core melt).

Regarding severe accident phenomena analysis, the relevant phenomena taken into account in Level 2 PSA and the main ways to assess their impact on the containment were presented. For some phenomena, a pressure load is determined by SA calculation codes and compared to the containment fragility curve. For others, the assessment relies on expert judgment, based on supporting studies and/or publications.

As regards to HRA, a new methodology called HAMSTER was presented. It allows to take into account multiple interactions occurring in a complex 'operating system' (e.g. operating crew in main control room, field operators, emergency teams, operating procedures). This method is applicable to Level 1 PSA and to Level 2 PSA, and to internal events PSA as well as internal or external hazards. Although the methodology and resulting assessment are strongly linked to EDF procedures and organization, the method can be easily adapted for other operating and crisis organizations.

Some specific Level 2 PSA considerations were discussed in the HRA framework. In general, no dependency factors are considered between human factor events (HFE) required before (i.e. as part of EOPs in Level 1 PSA) and after core damage (i.e. as part of SAMG in Level 2 PSA), as an important change of plant operation configuration occurs when entering SAMGs.

After the accident at the Fukushima Daiichi NPP, EDF created the FARN ('Force d'Action Rapide du Nucléaire', 300 people spread over five sites in France) to ensure that robust equipment and qualified external human resources are available at all times, in particular to mitigate any radioactive releases into the environment in the event of a severe accident. A HRA methodology is still under development for assessing HFE crediting portable equipment (brought by FARN or already available on the site). Nevertheless, a bounding assessment is currently being undertaken, taking into account the available time to perform the dedicated action; the adverse conditions on site and nearby (climatic, road access, radiation, etc.); and the need to potentially perform numerous actions (and the prioritization strategy applied).

A very simplified Level 2 PSA for SFP is carried out due to the absence of containment (the SFP is located outside the reactor building) and mitigation means. Thus, the release frequencies come from a direct allocation from CDF (differentiating accidental draining and loss of cooling accidents).

In the framework of the 4th 900 MW(e) PSR, many changes have been made in order to improve the safety level of these reactors. In terms of severe accident management, two areas of improvement have been identified: avoiding basemat failure (corium melt-through) and avoiding containment venting opening. This had led to the implementation of two main severe accident management features: an ex-vessel corium stabilization system (whose objective is to increase the surface of the spreading area of the corium by connecting the reactor pit and an adjacent room) and a dedicated containment heat removal system designed for severe accident conditions. Modifications implemented in the framework of the 4th 900 MW(e) PSR have resulted in a significant reduction in the important release frequency.

For operating plants, strategies for dealing with core melt and design of additional safety features are developed on a deterministic basis. Level 2 PSA is used to check the efficiency (from a probabilistic point of view) of these additional safety features, and to identify possible design or operations improvements.

Any baseline internal event PSA developed at EDF addresses a single unit on a nuclear site and some multi-unit aspects may be included in a simplified way in baseline external hazards PSAs. However, a more refined sensitivity analysis to multi-unit aspects has been carried out based on a practical approach of which an overview is given in the presentation. The analysis, focused on loss of offsite power and loss of ultimate heat sink events for twin units, was limited to the Level 1 PSA. The application of this practical approach showed that, for EDF PWRs, the risk increase and PSA insights considering multi-unit aspects are limited and thus there is no need to go further than the actual baseline single-unit PSA model.

In general terms, risk increase due to multi-unit aspects strongly depends on the design characteristics of the plants that are present on the site. Plant designs with many individual unit-specific mitigation means and few individual shared systems are less impacted by multi-unit aspects than plant designs with many individual shared systems. Thus, for EDF, the need for a full scope multi-unit PSA, which may be time consuming (because it involves the development of a large PSA model including all units), needs to be examined carefully. In some cases, a practical approach to take into account multi-unit aspects may appear to be sufficient to capture the main probabilistic insights.

### 2.5.7. Level 2 probabilistic safety assessment for spent fuel pool of the Armenian nuclear power plant, Armenia

This presentation provided the approach used to implement Level 2 PSA for the Armenian NPP SFP. The presentation consisted of three parts: the background and objectives, Level 1 and Level 2 PSA for the Armenian NPP SFP. The Armenian NPP is a VVER-440 type reactor where the SFPs are located outside the containment. This presentation was intended to present the approaches and results of the internal event PSA study performed for the Armenian NPP SFP.

The initial scope of the SFP probabilistic analysis study was internal events for both SFPs at full power operation. Proceeding from that circumstance that one of the study objectives is to provide the Armenian Nuclear Regulatory Authority (ANRA) with appropriate technical background for its decision making. The Nuclear and Radiation Safety Center (NRSC), a technical support organization of ANRA, decided to extend the scope of the project and include operating states with one tier of stacked fuel in the SFP and give a qualitative assessment of the release of radioactive material to the environment.

The presentation covered the major tasks of the internal event Level 1 PSA performed for the SFP of the Armenian NPP: internal initiating event analysis, accident sequence analysis, supporting deterministic analysis, system analysis, HRA, data analysis, dependency analysis, and the results. Level 1 PSA performed for the Armenian NPP SFPs showed that the main contribution to the fuel damage frequency (FDF) of Unit 1 SFP is mechanical damage of the limited number of fuel assemblies, meanwhile for Unit 2 the main contributions to the FDF are the human induced loss of coolant accident and mechanical damage of the limited number of fuel assemblies.

Regarding the Level 2 PSA the objective of the project was to give the qualitative assessment of the release of radioactive material to the environment based on Level 1 PSA results. For that, the results of Level 1 PSA were grouped based on the amount of fuel damaged and time to damage. The fuel damage states were grouped based on the potential amount of the releases: 'negligible' and large; the evaporation was not considered as a release. The frequencies of large releases from the SFP of Unit 1 and Unit 2 are 5.82 x $10^{-7}$ per reactor-year and 5.82 x $10^{-7}$ per reactor-year respectively, meanwhile for 'negligible' releases from the SFP of Unit 1 and Unit 2 are 3.26 x $10^{-5}$ per reactor-year and 1.13 x $10^{-4}$ per reactor-year respectively.

### 2.5.8. Level 2 probabilistic safety assessment of Slovak nuclear power plants, Slovakia

The defence in depth concept for NPPs involves deploying multiple levels and layers of protection to ensure safety. These include physical barriers to ensure the achievement of fundamental safety functions, and emergency response measures. The containment, as the final barrier, has to withstand loads from various severe accidents and prevent significant radioactive releases into the environment. The Level 2 PSA studies of the VVER plants in Slovakia were prepared for internal events, internal hazards and external hazards to, as stated in Ref. [47]:

- "To identify the ways in which radioactive releases from the plant can occur following the core damage;
- "To calculate the magnitudes and frequency of the release;
- "To provide insights into the plant behaviour during a severe accident;
- "To provide a framework for understanding containment failure modes, the impact of the phenomena that could occur during and following core damage and have the potential to challenge the integrity of the containment;
- "To support the severe accident management and development of guidelines."

This presentation described the main results and the lessons learned from the Level 2 PSA studies of the VVER 440 type reactors in Slovakia.

Based on the analyses performed within the Level 1 PSA, these external events were included into the Level 2 PSA: earthquake, extreme wind, tornado, extreme snow, extreme rain, extremely high air temperature, extremely low air temperature, icing and lighting. Human induced external events are not included in the Level 2 PSA models. Frequency of aircraft crash was determined to be less than 1.0 x $10^{-7}$ per year. Thus, the aircraft crash is excluded from further analysis. Similarly, influence of neighbouring industry and other external influences were estimated and excluded from further analysis.

The results of the analysis are compared with the probabilistic safety criteria defined for the Slovak plant:

- LERF < 1.0 x $10^{-5}$ per year plant in operation (including external hazards);
- LERF < 1.0 x $10^{-6}$ per year new plant (including external hazards).

The results are used to determine whether there are any weaknesses in the design and operation of the plant. Where such weaknesses are identified, considerations are made to reduce the risk from the severe accident. This typically includes the additional safety systems to provide protection for some of the adverse consequences of a severe accident. This is the case for Slovak NPPs, where the systems were completed and the guidelines for severe accident management were implemented. PSA results evaluated the benefits of these safety measures in terms of risk reduction.

The total LERF for internal events, internal hazards and external hazards were determined to be:

- Less than 1.0 x $10^{-5}$ per year for operating plants (Unit 1 and 2 of the Mochovce plant and Unit 3 and 4 for the Bohunice plant);
- Less than 1.0 x $10^{-6}$ per year for plants under construction (Unit 3 and 4 of the Mochovce plant).

These values indicate that results of large early releases are in compliance with established probabilistic safety criteria defined for the Slovak nuclear power plants.

#### 2.5.9. Updates to PWROG large early release frequency probabilistic risk assessment modelling methods, United States of America

##### *2.5.9.1. Current PWROG simplified Level 2 PRA methods*

As a basis, current LERF methods were described:

- Containment bypass: Covers steam generator tube rupture (SGTR) and interfacing system loss of coolant accident initiating events. Scrubbing credit is generally limited to isolated SGTR scenarios only.
- Containment isolation failure: Have a methodology for calculating a probability of having a pre-existing large containment leak, which is based on extrapolating results from the Integrated Leak Rate Test interval.
- Intentional reactor coolant system (RCS) depressurization mechanisms: RCS is something that interfaces with SAMG and is relatively limited. RCS was used because SAMG was separated with Westinghouse and Combustion Engineering. Have a human error probability (HEP) of either 0.1 or 0.5.
- Consequential SGTR: Have two different models, pressure and thermally induced SGTR. Both models include conditional SGTR probabilities which is derived from Ref. [48].
- Hydrogen combustion: Calculate peak pressures based on simplified calculations from Ref. [49] and assume adiabatic isochoric complete combustion conditions.
- Direct containment heating based on Ref. [50] for Westinghouse NSSS plants and based on Ref. [51] for Combustion Engineering NSSS plants;
- Ex-vessel steam explosions considered based on Ref. [29].

##### *2.5.9.2. PWROG SAMG development and impact on Level 2 PSA*

PWROG SAMG for the US plants were completed in February 2016 and SAMG for international plants were completed in December 2016. A programme to integrate the international reference plant into the PWROG SAMG started in April 2021.

The impact of SAMG updates on Level 2 PRA operator actions:

- Includes procedure-type guidance for the main control room operators for early priority actions including RCS depressurization;

- Integration of FLEX and other portable equipment;
- Improved guidance for containment venting.

#### 2.5.9.3. *Updates to PWROG LERF models and methods (2021–2022)*

The PWROG expanded the applicability of the PWROG LERF models and methods for:

- Westinghouse, Combustion Engineering and Babcock and Wilcox NSSS designs;
- Large dry containments and ice condenser containments;
- Methods expanded to include plants with passive autocatalytic recombiners (PARs) and containment filtered vents.

Improvements in the PWROG SAMG allow for improved reliability of post-core damage operator actions to be quantified:

- Lower human error probabilities for intentional RCS depressurization.
- Improved reliability for containment venting. Containment venting for pressure control is unlikely to impact LERF sequences; however, this may need to be confirmed with plant specific analysis, especially for plants with lower design pressure.
- Risk insights from key post-core damage operator actions can help inform future revisions of PWROG SAMG.

#### 2.5.9.4. *Incorporation of modern research*

Insights from three recent publications will be included in the new PWROG LERF methodology: NRC thermally induced steam generator tube rupture study [52], state of the art reactor consequence analysis (SOARCA) [53], EPRI technical basis report (TBR) [54].

Recent studies are expected to reduce conservatism in LERF PRAs in the following areas:

- Reduce the assumed dose consequences of isolated and un-isolated SGTR core damage sequences;
- Reduce uncertainty in treatment of thermally induced SGTRs.

### 2.5.10. Summary of Session V

The experience from nuclear industry facilities in several Member States was presented. During the section discussions the following highlights were made:

- As stated in Ref. [45], "Severe accident management and mitigation measures could be reduced by a factor of at least ten for the probability of large off-site releases requiring short-term off-site response." For modern designs, this seems unrealistic (e.g. plants with the estimated CDF to be in the range of 1 x $10^{-7}$ per year, with the risk profile being dominated by irreducible contributors, it needs to be recognized that providing a low LERF/CDF ratio is rather impossible).
- The requirement to have balanced design is not applicable for modern designs where we have low CDF, because few contributors will have the most impact (up to 99%), due to the lack of knowledge of these particular issues (e.g. vessel rupture).

The results of the discussions were considered as appropriate in the revised version, which are summarized below:

- There is a need to consider in the source term calculations other sources of potential release.
- The Level 2 PSA methodology aims to be technology inclusive rather than technology neutral or technology specific.
- There were suggestions in relation to the definition of risk metrics and probabilistic safety goals for Level 2 PSA. It is understood that these suggestions aim at supporting Member States to define risk metrics and probabilistic safety goals for Level 2 PSA in their national regulations with the objective to facilitate independent peer reviews and harmonization of regulatory practices rather than provide quantitative values.
- There is a need to provide further recommendations in relation to integral and separate approaches for development Level 1 PSA and Level 2 PSA keeping in mind the latest developments for new NPPs and for advanced NPPs for which the Level 1 PSA risk metrics are less relevant than those of Level 2 PSA.
- There is a need for recommendations related to the reassessment of the list of internal and external hazards and consideration to address their combination of hazards in Level 2 PSA.
- There is a need to clarify the objective of recommendations related to the development of Level 2 PSA for multi-unit site considering the lack of consensus and general practice in Member States.
- There were suggestions to provide further recommendations related to the modelling of HRA in Level 2 PSA.
- There is a need for recommendations related to the development of Level 2 PSA for SFP.
- There is a need for examples of specific applications of Level 2 PSA, including the importance of risk informed applications that provide a balance between conservatism and best estimate assumptions.
- Examples of recent results of research programmes in relation to severe accident strategies need to be considered.

During the discussion, the definition of the general objective for the development of Level 2 PSA was discussed: to verify the robustness of the severe accident management strategies to implement effective protection of the population and the environment against severe accidents, for which particularly high quality Level 2 PSAs are required. For this purpose, meeting numerical safety goals is not primarily required, but a comprehensive understanding of assumptions and severe accident phenomena relevant to the NPP technology and design. However, Level 2 PSA results could contribute to the demonstration of the 'practical elimination' concept for plant event sequences that could lead to an early radioactive release or a large radioactive release. It was also highlighted that Level 2 PSA can be used in the process of verifying the implementation of possible severe accident strategies when hazards occur.

# 3. SUMMARY OF THE TECHNICAL MEETING

## 3.1. SUMMARY

The experience gained during the technical meeting was identified and grouped in the most relevant area for deeper processing and inclusion in SSG-4 (Rev. 1) [3]. The conclusions are presented below.

### 3.1.1. Technology neutral vs technology specific and inclusive

SSG-4 [2] provides high level recommendations, and the proposed methodology of Level 2 PSA is applicable to all reactor technologies. The understanding of the term 'technology neutral' will apply only to a general probabilistic safety assessment methodology, as general quantitative risk assessment method, where no specifics related to reactor technology will be presented. However, a 'technology inclusive' term is proposed since the recommendations recognize specific aspects related only to nuclear power plants such as reactor core, reactor containment, SFP and release categories. However, to complement the technology inclusive methodology, it is understood that the phenomena associated with severe accidents are technology and design specific, therefore the recommendations provided in SSG-4 [2], which are primarily based on experience with water cooled reactor technologies, might need to apply engineering judgement when adapted to other reactor technologies. SSG-4 (Rev. 1) [3] aims at presenting recommendations as technology inclusive rather than technology neutral.

### 3.1.2. Definition of risk metrics and safety goals for Level 2 probabilistic safety assessment

Currently Members States have different definitions of probabilistic safety goals for Level 2 PSA. For example, there was no consensus in the definition of terms such as 'large' and 'early' release. In that regard, it is clear that risk metrics LERF and LRF have to be defined based on measurable values: 'magnitude', 'time' and 'frequency', and include relevant information about the radioactive substances that could be released in terms of chemical elements and forms, as well as either absolute values or percentage of core inventory of specific radionuclides. Further suggestions related to the definition of previous aspects are provided in Section 2.2.7.

Furthermore, the definition of probabilistic safety goals for Level 2 PSA needs to consider the new design features proposed by new NPP and advanced NPP designs. For instance, many advanced NPP designs propose a lower reactor thermal power, facilitating residual heat removal through passive safety systems and natural circulation, together with other design safety features aimed at reinforcing the confinement function. The incorporation of those design safety features in their designs results in achieving lower LRF or LERF values compared to existing probabilistic safety goals for NPPs currently in operation (e.g. from Ref. [45]). This, therefore raises the question about their applicability for new NPP designs. Another aspect to be considered is related to the balanced risk profile of the NPP design, which for some advanced NPP designs might be rather difficult to achieve. Indeed, the results of Level 2 PSA will reflect the advantages of incorporating design safety features in those advanced NPP designs, leading the risk picture to be dominated by single initiating event or hazard which could have a major contribution to the overall CDF/LRF (e.g. reactor vessel rupture or some seismic events). Then, the difficulty to propose alternative design options and reinforcements to reduce the impact of such internal initiating events or external hazards with a dominant contribution to the CDF/LRF means, due to the high uncertainties associated with these events, has to be considered. Therefore, it might not be advisable to reduce the probabilistic safety goals for Level 2 PSA for advanced NPPs with regard to generation III and III+ NPPs designs one decade lower. Instead, further recommendations related to the validation of methods used for the development of Level 2 PSA, the assumptions considered as well as the analysis of the associated uncertainties to the Level 2 PSA results rather than meeting the quantitative value seem more important. SSG-4 (Rev. 1) [3], together with the new edition of the ASME standard for Level 2 PSA, aim to achieve such an objective.

Considering the above discussions, and that IAEA safety standards are based on a consensus reached over the best international practices, it is challenging to provide recommendations in SSG-4 (Rev. 1) [3] including quantitative values related to Level 2 PSA probabilistic safety goals. Instead, SSG-4 (Rev. 1) [3] provides recommendations on aspects to be considered when national regulations define Level 2 PSA probabilistic safety goals allowing for a clear understanding among regulatory bodies, designers and operating organizations, facilitating independent peer reviews and contributing to the harmonization of regulatory frameworks among Member States.

#### 3.1.3. Level 2 probabilistic safety assessment for the other sources of potential radioactive releases

The need for quantification in the source term calculation of potential radioactive releases from all sources on the site as the outcome of Level 2 PSA has been confirmed during the meeting discussions. Therefore, given the significant quantity of radioactive substances located in the SFP and the potential consequences for the environment and the population in case of significant damage of that fuel, the development of Level 2 PSA for the SFP is an important task. However, there are several aspects to be considered. The first point is related to the criteria corresponding to the fuel damage in the SFP defined in the Level 1 PSA. Indeed, several Member States have mentioned different criteria for defining fuel damage in the SFP for the Level 1 PSA, such as boiling of the SFP, fuel uncovering, and fuel damage. Another aspect is related to the location of the SFP in a structure capable to ensure an effective confinement function, such as inside or outside reactor building. The last aspect to consider is related to the combined impact on the energy and mass released in the containment when combined severe accidents happen in the reactor core and the SFP which is located inside the reactor containment building. Therefore, the difference between NPP designs makes difficult to provide common recommendations in the development of Level 2 PSA.

In relation to the consideration of other potential radioactive releases, besides reactor core and the SFP, in the source term calculation, the discussions agreed that there is a need to propose recommendations in SSG-4 (Rev. 1) [3]. First, all types and forms of radioactive substances stored on the NPP site need to be identified. Currently, the Level 2 PSA approach allows the consideration of most radioactive substances in the source term calculation, however there is an issue related to sources (e.g. radioactive waste) stored outside an effective structure ensuring the confinement function. The potential impact of those sources can be quantified in the overall source term calculation for the site, however the Level 2 PSA methodology, as used for the reactor core, could be significantly simplified, because incidents on such sources could lead to a direct release to the environment. Therefore, despite the lack of regulatory consensus on this specific topic among Member States, the meeting discussions agreed that a conservative approach will be acceptable.

Additionally, there is an issue related to risk aggregation for Level 2 PSA results when considering the reactor core, SFPs, and other sources of radioactive release, which needs to be solved for the risk informed decision making application.

#### 3.1.4. Reassessment of the list of internal and external hazards and considerations to address their combination of hazards in Level 2 probabilistic safety assessment

The accident at the Fukushima Daiichi NPP has reinforced the need to consider the combination of internal hazards and external hazards in recent PSA developments. The process for assessing internal hazards, external hazards and their combinations is considered in the development of

Level 1 PSA. In many developments of Level 2 PSA in Member States, the list of internal hazards, external hazards and their combinations consider only those which were not screened out in Level 1 PSA. However, during the technical meeting the need to reassess the screening process defining the list of internal hazards, external hazards and their combinations to be considered for a Level 2 PSA was discussed. The reassessment of the screening process needs to consider both the consequence and the frequency of the combination of hazards. The main reasons justifying the reassessment were related to the difference in mission times and equipment modelled in both Level 1 PSA and Level 2 PSA. Therefore, the potential impact of hazards on the containment integrity as well as the dependent failures, which can be induced by hazard combinations, need to be considered in the development of Level 2 PSA, if those aspects have not yet been considered in the Level 1 PSA.

### 3.1.5. Considerations related to multi-unit sites for Level 2 probabilistic safety assessment

The IAEA provides a series of examples and good practices related to the methodology for development of multi-unit PSA. However, there is no common consensus or experience on the development of this practice in Member States. In particular, one aspect of the multi-unit methodology requiring clarity is related to the definition and understanding of probabilistic safety goals for multi-unit Level 2 PSA. However, it is understood that the methodology for multi-unit Level 2 PSA is similar to that for Level 1 PSA, which has to consider all possible dependencies on the site mainly related to the site resources such as human, equipment, organization and possible common caused failures.

For example, common cause failures need to be considered following a realistic approach on those pieces of equipment that could be affected by the initiating event such as batteries, water tanks or emergency diesel generators. However, the introduction of those common cause failures in the multi-unit model depends on the assumptions considered for the single unit model and on the assumptions related to the extrapolation to the other units nearby.

Considering the lack of consensus on the practice for development Level 2 PSA for a multi-unit site, it was agreed that there could be recommendations provided aiming at harmonization of the practice in SSG-4 (Rev. 1) [3]. However, those recommendations need to specify that they only apply to those Member States where there is a regulatory requirement compelling the development of such PSA developments. Otherwise, a simplified approach could be practical to assess the risk level for a site where more than one unit is installed. In SSG-4 (Rev. 1) [3], this recommendation has been taken into consideration and allows for flexibility in relation to multi-unit Level 2 PSA.

### 3.1.6. Further explanation on integral vs separate models

Two approaches are established in relation to the development of Level 1 PSA and Level 2 PSA studies in Member States. Those approaches basically define the transition of information from a Level 1 PSA model to a Level 2 PSA model, which are either a separated model approach or an integrated model approach.

Both approaches have their own features and specific advantages and disadvantages, see Section 2.2.7. In addition, it is important to highlight that most recent developments of PSA studies either for Generation III NPP or for advanced NPP designs follow the integrated approach. The main reasons for considering the integrated approach rather than the separate approach are related to regulatory requirements, the reliability of design safety features in those

NPP designs and subsequently the results obtained in Level 1 PSA (see Section 2.2.7). On the contrary, for NPPs in current operation, the separated approach has been followed for the development of Level 1 PSA and Level 2 PSA and it is still widely used in many Member States. SSG-4 (Rev. 1) [3] will provide recommendations in relation to the use of both possible approaches.

### 3.1.7. Human reliability analysis in Level 2 probabilistic safety assessment

Human actions have great impact in PSA and the meeting discussions covered them in the context of development of Level 2 PSA. In order to assess the effectiveness of human actions, HRA methods are conducted as part of the development of PSA studies. Since current HRA methods for Level 1 PSA might not be fully applicable for Level 2 PSA, several adaptations to existing HRA methods or new HRA methods have been developed. These HRA methods need to reassess assumptions to quantify HEP made in Level 1 PSA, since there might be changes such as due to increased stress factor related to the management of severe accident situations. One aspect discussed was related to the integration of the SAMG in the development of Level 2 PSA, considering that many Level 2 PSA studies did not consider SAMG from the beginning. Another aspect to be considered was related to the inclusion of additional safety features for mitigating design extension conditions with core melting and the deployment of non-permanent equipment to ensure electrical power supply, water supply, compressed air and batteries. Therefore, the integration of previous aspects in the development of Level 2 PSA requires the evolution of HRA methods. SSG-4 (Rev. 1) [3] will provide recommendations related to HRA methods to be considered in the development of Level 2 PSA.

### 3.1.8. Applications of Level 2 probabilistic safety assessment

Several practical applications of Level 2 PSA results were presented during the technical meeting. Most of them are mainly for risk monitoring application, which allow to monitor the level of risk change (i.e. delta LRF or LERF) as part of the risk informed decision making process (e.g. in relation to design modifications, change in technical specification, changes on NPP configuration or maintenance). Nonetheless the practical application of Level 2 PSA results is not widely used in Member States due to the high level of uncertainties associated with their results. SSG-4 (Rev. 1) [3] will provide recommendations encouraging the use of Level 2 PSA studies in risk informed decision making process applications. Other examples related to the application of risk informed decision making process are provided in Ref. [9].

Another important application of Level 2 PSA results is related to the demonstration of the practical elimination concept. The practical elimination concept was developed in the 1990s by the French and German advisory groups for the construction of new NPPs equipped with PWR and later integrated in Ref. [55]. The idea of practical elimination complements the defence in depth approach by ensuring that adequate, sufficient and robust safety provisions will be considered in the NPP design allowing to effectively exclude the occurrence of those plant event sequences that could lead to unacceptable consequences to the population and the environment in terms of length of time and area. Therefore, the practical elimination concept is applied to those plant event sequences that could lead to an early radioactive release or a large radioactive release. The demonstration of the application of this concept considers the physical impossibility of the event to happen or to ensure with high degree of confidence that the event is extremely unlikely to happen. While the use of the first option for the demonstration of the practical elimination concept seems more deterministic, it might be only applicable to few cases. There was agreement during the technical meeting discussions that the application of Level 2 PSA studies could contribute to the demonstration of the second option by providing insights related to the probabilistic terms used such as 'high degree of confidence' and 'extremely unlikely'. However, there is consensus that the demonstration of the application of the practical elimination concept cannot rely only on meeting probabilistic values.

### 3.1.9. Examples of research programmes

The development of Level 2 PSA methodology is continuing to evolve and being enhanced by the incorporation of knowledge related to research activities on severe accident phenomena. Relevant research and development programmes are conducted in several Member States and during the technical meeting some examples were presented. The most important areas of research and development are the following:

- In-vessel or ex-vessel phenomenology of severe accidents;
- Dynamic of accident progression;
- Development of calculation codes for improving the realism of modelling;
- Analysis of uncertainties;
- Human reliability analysis;
- Assessment of release consequences;
- Development of SAMG;
- Risk aggregation of Level 2 PSA results.

## 3.2. PATH FORWARD

The technical meeting demonstrated the accumulated practical experience and significant level of efforts in IAEA Member States on Level 2 PSA approaches. While these efforts are noteworthy, they have demonstrated the methodological challenges associated with this area. Therefore, continued coordination at the international level is suggested by the meeting participants to coordinate these efforts in order to develop optimal approaches addressing the topic of 'Level 2 PSA' and subsequent more deep and realistic safety assessment of nuclear installations.

The contributions presented during the meeting were devoted to new methodological developments and implementations of interesting aspects of Level 2 PSA approaches, projects with implementation such activities and expected challenges for future analyses.

The selection of the indicated topics for the technical meeting discussion was aimed at collecting the experience on the development and application of Level 2 PSA for NPPs covering a wide range of tasks (e.g. methods, approaches and assumptions). The technical meeting discussions also highlighted their influence in the Level 2 PSA results, advantages and disadvantages, and confirm the need to apply the realism expected for PSA analysis.

This technical meeting was concurrent with completed and ongoing activities in IAEA:

- Revision of SSG-3;
- Development of a TECDOC on "Advanced Probabilistic Safety Assessment (PSA) Approaches and Applications for Nuclear Power Plants" that contains general information corresponding to additional approaches used in PSA;
- Development of new Safety Reports Series on Human Reliability Analysis;
- Development of a TECDOC on "Risk aggregation";
- MUPSA Projects, including CRP projects and development new TECDOC;

- Safety assessment for SMR.

Collected Member States experience on Level 2 PSA and output from the technical meeting will be used as key inputs to the revision of Safety Guide SSG-4.

## REFERENCES


[1] INTERNATIONAL ATOMIC ENERGY AGENCY, Development and Application of Level 1 Probabilistic Safety Assessment for Nuclear Power Plants, IAEA Safety Standards Series No. SSG-3 (Rev. 1), IAEA, Vienna (2024), https://doi.org/10.61092/iaea.3ezv-lp49.

[2] INTERNATIONAL ATOMIC ENERGY AGENCY, Development and Application of Level 2 Probabilistic Safety Assessment for Nuclear Power Plants, IAEA Safety Standards Series No. SSG-4, IAEA, Vienna (2010).

[3] INTERNATIONAL ATOMIC ENERGY AGENCY, Development and Application of Level 2 Probabilistic Safety Assessment for Nuclear Power Plants, IAEA Safety Standards Series No. SSG-4 (Rev. 1), IAEA, Vienna (in preparation).

[4] INTERNATIONAL ATOMIC ENERGY AGENCY, Safety Assessment for Facilities and Activities, IAEA General Safety Requirements GSR Part 4 (Rev. 1), IAEA, Vienna (2016).

[5] INTERNATIONAL ATOMIC ENERGY AGENCY, Safety of Nuclear Power Plants: Design, IAEA Specific Safety Requirements No. SSR-2/1 (Rev. 1), IAEA, Vienna (2016).

[6] INTERNATIONAL ATOMIC ENERGY AGENCY, Safety of Nuclear Power Plants: Commissioning and Operation, IAEA Specific Safety Requirements No. SSR-2/2 (Rev. 1), IAEA, Vienna (2016).

[7] INTERNATIONAL ATOMIC ENERGY AGENCY, Evaluation of Seismic Safety for Nuclear Installations, IAEA Safety Standards Series No. SSG-89, IAEA, Vienna (2024), https://doi.org/10.61092/iaea.jtrh-4y3z

[8] INTERNATIONAL ATOMIC ENERGY AGENCY, Multi-unit Probabilistic Safety Assessment, Safety Reports Series No. 110, IAEA, Vienna (2023)

[9] INTERNATIONAL ATOMIC ENERGY AGENCY, Considerations on Performing Integrated Risk Informed Decision Making, IAEA-TECDOC-1909, IAEA, Vienna (2020).

[10] INTERNATIONAL ATOMIC ENERGY AGENCY, Probabilistic Safety Assessment for Seismic Events, IAEA-TECDOC-1937, IAEA, Vienna (2020).

[11] INTERNATIONAL ATOMIC ENERGY AGENCY, Attributes of Full Scope Level 1 Probabilistic Safety Assessment (PSA) for Applications in Nuclear Power Plants, IAEA-TECDOC-1804, IAEA, Vienna (2016).

[12] SCIENTIFIC AND ENGINEERING CENTRE FOR NUCLEAR AND RADIATION SAFETY, Federal Rules and Regulations in the Field of Nuclear Energy Use "General Safety Provisions for Nuclear Power Plants", NP-001-15 (2016).

[13] SCIENTIFIC AND ENGINEERING CENTRE FOR NUCLEAR AND RADIATION SAFETY, Federal Rules and Regulations in the Field of Nuclear Energy Use "Basic

Requirements to Probabilistic Safety Analysis of a Nuclear Power Plant Unit", NP-095-15, Moscow (2015).

[14] SCIENTIFIC AND ENGINEERING CENTRE FOR NUCLEAR AND RADIATION SAFETY, "Recommendations on the development of Level 2 probabilistic safety analysis for NPP unit", RB-044-18, Moscow (2018).

[15] Ministry of Economic Affairs and Employment, Finland, Nuclear Energy Decree, Section 22b (1001/2017) 12.2.1988/161.

[16] THE RADIATION AND NUCLEAR SAFETY AUTHORITY (STUK), Probabilistic risk assessment and risk management of a nuclear power plant, YVL A.7 (2019), https://www.stuklex.fi/en/ohje/YVLA-7

[17] ARGENTINE NUCLEAR REGULATORY AUTHORITY, Memoria Técnica ARN 2021, Probabilistic Approach for Nuclear Safety Regulation in Argentina, pp, 101-110, https://www.argentina.gob.ar/sites/default/files/mt_arn_2021.pdf

[18] CANADIAN NUCLEAR SAFETY COMMISSION, Probabilistic Safety Assessment (PSA) for Nuclear Power Plants, regulatory document REGDOC-2.4.2, (May 2014).

[19] CANADIAN STANDARD ASSOCIATION Group, Probabilistic Safety Assessment for Nuclear Power Plant, CSA N290.17-17 (2017).

[20] CANADIAN NUCLEAR SAFETY COMMISSION, Regulatory Document REGDOC-2.5.2, Physical Design, Design of Reactor Facilities: Nuclear Power Plants, Ottawa, ON, Canada (May 2014).

[21] INTERNATIONAL NUCLEAR SAFETY ADVISORY GROUP, Basic Safety Principles for Nuclear Power Plants, 75-INSAG-3, IAEA, Vienna (1988).

[22] INTERNATIONAL ATOMIC ENERGY AGENCY, Review of Probabilistic Safety Assessments by Regulatory Bodies, IAEA Safety Report Series No. 25, IAEA, Vienna (2002).

[23] INTERNATIONAL ATOMIC ENERGY AGENCY, Regulatory Review of Probabilistic Safety Assessment (PSA) Level 2, IAEA-TECDOC-1229, IAEA, Vienna (2001).

[24] Modular Accident Analysis Program for CANDU reactor (MAAP5-CANDU), version 5.00a. EPRI, Palo Alto, CA: 2018. 3002012973, 3002012974.

[25] INTERNATIONAL COMMISSION ON RADIOLOGICAL PROTECTION, The 2007 Recommendations of the International Commission on Radiological Protection, Publication 103, Elsevier (2007).

[26] EUROPEAN COMMISSION, FOOD AND AGRICULTURE ORGANIZATION OF THE UNITED NATIONS, INTERNATIONAL ATOMIC ENERGY AGENCY, INTERNATIONAL LABOUR ORGANIZATION, OECD NUCLEAR ENERGY AGENCY, PAN AMERICAN HEALTH ORGANIZATION, UNITED NATIONS ENVIRONMENT PROGRAMME, WORLD HEALTH ORGANIZATION, Radiation

Protection and Safety of Radiation Sources: International Basic Safety Standards, IAEA Safety Standards Series No. GSR Part 3, IAEA, Vienna (2014).

[27] FOOD AND AGRICULTURE ORGANIZATION OF THE UNITED NATIONS, INTERNATIONAL ATOMIC ENERGY AGENCY, INTERNATIONAL CIVIL AVIATION ORGANIZATION, INTERNATIONAL LABOUR ORGANIZATION, INTERNATIONAL MARITIME ORGANIZATION, INTERPOL, OECD NUCLEAR ENERGY AGENCY, PAN AMERICAN HEALTH ORGANIZATION, PREPARATORY COMMISSION FOR THE COMPREHENSIVE NUCLEAR-TEST-BAN TREATY ORGANIZATION, UNITED NATIONS ENVIRONMENT PROGRAMME, UNITED NATIONS OFFICE FOR THE COORDINATION OF HUMANITARIAN AFFAIRS, WORLD HEALTH ORGANIZATION, WORLD METEOROLOGICAL ORGANIZATION, Preparedness and Response for a Nuclear or Radiological Emergency, IAEA Safety Standards Series No. GSR Part 7, IAEA, Vienna (2015).

[28] Griesmeyer, J. M., and Smith, L. N. A reference manual for the Event Progression Analysis Code (EVNTRE). United States: N. p., 1989. Web. doi:10.2172/5687655.

[29] U.S. NUCLEAR REGULATORY COMMISSION, An Approach for Estimating the Frequencies of Various Containment Failure Modes and Bypass Events, NUREG/CR 6595, Rev. 1, Brookhaven National Laboratory, Washington, DC (2004).

[30] INSTITUTE FOR RADIOLOGICAL PROTECTION AND NUCLEAR SAFETY, Best Practices guidelines for Level 2 PSA development and applications, Volume 1- General, ASAMPSA2, IRSN-PSN/RES/SAG 2013-0177 (2013).

[31] INTERNATIONAL ATOMIC ENERGY AGENCY, Advances in Small Modular Reactor Technology Developments, a Supplement to IAEA Advanced Reactors Information System (ARIS), IAEA Booklet, 2020 Edition, IAEA, Vienna (2020).

[32] Western European Nuclear Regulators Association, Report 'Safety of new NPP designs', Study by Reactor Harmonization Working Group RHWG, (2013).

[33] U.S. NUCLEAR REGULATORY COMMISSION, Shutdown and Low-Power Operation at Commercial Nuclear Power Plants in the United States, NUREG-1449, Washington, DC (1993);

[34] INTERNATIONAL ATOMIC ENERGY AGENCY, PSA for the Shutdown Mode for Nuclear Power Plants, IAEA TECDOC-751, IAEA, Vienna (1994).

[35] INTERNATIONAL ATOMIC ENERGY AGENCY, Use of PSA Level 2 Analysis for Improving Containment Performance, IAEA TECDOC-1002, IAEA, Vienna (1998).

[36] Y. Kirimoto, et al., "The Human Reliability Analysis (HRA) Guide with Emphasis on Narratives", CREPI Report O18011 (2018)

[37] SCIENTIFIC AND ENGINEERING CENTRE FOR NUCLEAR AND RADIATION SAFETY, Recommendations for developing of an ultimate list of beyond design accidents which shall be considered in NPP design with VVER type reactors, RB-150-18, Moscow (2018).

[38] THE AMERICAN SOCIETY OF MECHANICAL ENGINEERS, Severe Accident Progression and Radiological Release (Level 2) PRA Standard for Nuclear Power Plant Applications for Light Water Reactors (LWRs), ASME/ANS RA-S-1.2-2014 (2014).

[39] THE AMERICAN SOCIETY OF MECHANICAL ENGINEERS, Standard for Level 1/Large Early Release Frequency Probabilistic Risk Assessment for Nuclear Power Plant Applications, ASME/ANS RA-Sa-2009 (2009).

[40] THE AMERICAN SOCIETY OF MECHANICAL ENGINEERS, Standard for Level 1/Large Early Release Frequency Probabilistic Risk Assessment for Nuclear Power Plant Applications, ASME/ANS RA-Sb 2013, New York (2017).

[41] U.S. NUCLEAR REGULATORY COMMISSION, An approach for using probabilistic risk assessment in risk-informed decisions on plant specific changes to the licensing basis, Regulatory Guide 1.174 (2009).

[42] AMERICAN SOCIETY OF MECHANICAL ENGINEERS, Standard for Severe Accident Progression and Radiological Release (Level 2) PRA Standard for Nuclear Power Plant Applications for Light Water Reactors (LWRs), ASME/ANS RA-S-1.2-2019, ASME, New York (in preparation).

[43] UNITED STATES NUCLEAR REGULATORY COMMISSION, Staff Requirements – SECY-98-144 – White Paper on Risk-Informed and Performance- Based Regulation, US NRC, Washington, DC (1999).

[44] UNITED STATES NUCLEAR REGULATORY COMMISSION, Feasibility Study for a Risk-Informed and Performance-Based Regulatory Structure for Future Plant Licensing – Main Report, NUREG-1860, Vol. 1, US NRC, Washington, DC (2007).

[45] INTERNATIONAL NUCLEAR SAFETY ADVISORY GROUP, Basic Safety Principles for Nuclear Power Plants 75-INSAG-3 Rev. 1, INSAG-12, IAEA, Vienna (1999).

[46] ROSTEKHNADZOR, Regulations on Applying Risk informed Methods in Making Regulatory Decisions Concerning NPPs, RB-101-16., Moscow (2016).

[47] NUCLEAR ENERGY AGENCY COMMITTEE ON THE SAFETY OF NUCLEAR INSTALLATIONS, , Use and development of probabilistic safety assessment CSNI WGRISK, NEA/CSNI/R(2007)12, NEA (2007).

[48] U.S. NUCLEAR REGULATORY COMMISSION, Risk Assessment of Severe Accident Induced Steam Generator Tube Rupture, NUREG-1570, USNRC (1998).

[49] U.S. NUCLEAR REGULATORY COMMISSION, PWR Dry Containment Issue Characterization, J. W., Yang, NUREG/CR-5567 (1990).

[50] U.S. NUCLEAR REGULATORY COMMISSION, Resolution of Direct Containment Heating Issue for all Westinghouse Plants with Large Dry Containments or Subatmospheric Containments, NUREG/CR 6338, Sandia National Laboratories (1996).

[51] U.S. NUCLEAR REGULATORY COMMISSION, Resolution of the Direct Containment Heating Issue for Combustion Engineering Plants and Babcock & Wilcox Plants, NUREG/CR 6475, Sandia National Laboratories (1998).

[52] U.S. NUCLEAR REGULATORY COMMISSION, Consequential SGTR Analysis for Westinghouse and Combustion Engineering Plants with Thermally Treated Alloy 600 and 690 Steam Generator Tubes, NUREG-2195, USNRC (2018).

[53] U.S. NUCLEAR REGULATORY COMMISSION, State-of-the-Art Reactor Consequence Analyses Project – Volume 2: Surry Integrated Analysis, NUREG/CR-7110, Volume 2, Revision 1, Sandia National Laboratories (2013).

[54] THE ELECTRIC POWER RESEARCH INSTITUTE, Severe Accident Management Guidance Technical Basis Report Volume 2: The Physics of Accident Progression, EPRI 1025295, EPRI (2012).

[55] INTERNATIONAL NUCLEAR SAFETY ADVISORY GROUP, Defence in Depth in Nuclear Safety, INSAG Series No. 10, IAEA, Vienna (1996).

## ABBREVIATIONS

| | |
|---|---|
| AOT | Allowed outage time |
| APET | Accident progression event trees |
| ATWS | Anticipated transient without scram |
| CDF | Core damage frequency |
| CET | Containment event tree |
| DSA | Deterministic safety assessment |
| EOP | Emergency operating procedures |
| FDF | Fuel damage frequency |
| HEP | Human error probability |
| HFE | Human factor events |
| HRA | Human reliability analysis |
| IE | Initiating event |
| IVR | In-vessel retention |
| LERF | Large early release frequency |
| LRF | Large release frequency |
| LWR | Light water reactor |
| NPP | Nuclear power plant |
| PDS | Plant damage states |
| POS | Plant operation state |
| PSA | Probabilistic safety assessment |
| PSR | Periodic safety review |
| RCS | Reactor coolant system |
| RPV | Reactor pressure vessel |
| SAMG | Severe accident management guideline |
| SET | System event tree |
| SFP | Spent fuel pool |
| SFR | Sodium cooled fast reactors |
| SGTR | Steam generator tube rupture |
| SSC | Structure, system, and component |

## ANNEX I. TECHNICAL MEETING PARTICIPANTS SUGGESTIONS FOR REVISION OF SSG-4

The suggestions from the participants in the technical meeting, summarised in Table I-1, have been examined in detail and implemented, as appropriate, in SSG-4 (Rev. 1), which was in preparation at the time of this publication.

TABLE 4. SUGGESTIONS FOR THE REVISION OF SSG-4.

| | Issue | Proposed change |
|---|---|---|
| 1 | **Terms and terminology**<br>The terminology varies from other IAEA publications.<br>SSG-4 does not define the terms and definitions used in the publication. | SSG-4 uses the terminology of superseded IAEA publications (e.g. beyond design basis accident). Need to check for compliance with current Safety Requirements.<br><br>To add a section to define terms and definitions such as large, early, mission time, release category, source term, etc. |
| 2 | SSG-4 does not aim to address essential points, such as:<br>• SFP;<br>• Multi-unit, and multi-sources;<br>• Radioactivity outside the reactor core. | To provide recommendations for considerations of the SFP including two conceptual design solutions: SFP inside and SFP outside containment.<br>To provide recommendations for potential accidents involving multiple reactor units and SFPs concurrently (multi-unit and multi-sources Level 2 PSA).<br>To provide recommendations regarding radioactive waste stored on the site (this might be rather general, e.g. assess maximum possible release and identify propagation paths). |
| 3 | Information provided in para. 1.7 and Section 5 is not sufficient to define release categories in terms of quantities of released radioactive products (these quantities are recommended to be assessed at the step of Source terms analyses).<br>SSG-4 does not clearly address what are the endpoints of the containment event tree and the attributes of release categories. | To revise Figure 1 and items (3) and (4) of para. 1.7 to give a clear and consistent picture of the process.<br>To combine Sections 5 and 6 to have a complete picture of the endpoint of containment event trees and provide an illustrative example of containment event tree in graphical form. |
| 4 | **Section 2 "PSA Project Management and Organization"**<br>Section 2 includes subsection 2.1 "Definition of the objectives of Level 2 PSA", but does not provide any information on recommended or practiced probabilistic safety goals or criteria associated with Level 2 PSA. | To add subsection "Level 2 PSA probabilistic safety goals or criteria", in which it is suggested to discuss various risk metrics used in Level 2 PSA in Member States, relative safety goal and criteria practiced in Member States, as well as general recommendations on Level 2 PSA risk metrics and safety goals (criteria) that needs to be provided. |
| 5 | **Section 3 "Identification of design aspects important to severe accidents and acquisition of information"**<br>The section does not contain recommendations related to information needed for Level 2 PSA for:<br>• Radioactive sources other than fuel in the reactor;<br>• Hazards other than internal events;<br>• Other modes of operation than operation at power. | To add recommendations addressing external hazards and their combination in corresponding sections that describe the related information needed for Level 2 PSA for other sources of radioactivity, for shutdown operation, including external and internal hazards. |

| | Issue | Proposed change |
|---|---|---|
| 6 | **Section 4 "Interface with Level 1 PSA: grouping of sequences"**<br>Does not really provide useful information. | To address the following aspects:<br>• Treatment of recoveries in the L1–L2 interface model;<br>• Enhancement of Level 1 PSA model to add features needed for Level 2 PSA (e.g. consideration of abandoned systems);<br>• Attributes important for Level 2 PSA for shutdown states (Paras 4.11 and 4.12 do not provide really useful information);<br>• Consideration of mission time in the L1–L2 interface model (24 hours used in Level 1 PSA might not be sufficient). |
| 7 | Paragraph 4.7. of SSG-4 states:<br><br>"For plant damage states with containment bypass, the main consideration should be the identification of attributes that are associated with attenuation of concentrations of radioactive material along the release pathway or affect the timing of release. This should include the type of initiating event, the status of the emergency core cooling system (including failure time) and whether the leak pathway is isolable after a period or whether it passes through water (e.g. steam generator inventory or flooded building)."<br><br>Implicitly this text can be understood that if bypath is localized or submerged releases are over, reactor vessel damage still might occur and all containment phenomena are still applicable. | To add discussion that bypass end states can be divided into bypass localized and non-localized. For localized bypass (if core melt already occurs) phenomena in the reactor and containment still need to be considered. |
| 8 | Paragraph 4.10. of SSG-4 states:<br><br>"In order to extend the scope of the Level 2 PSA to include internal and external hazards, their impact on systems necessary for mitigation of severe accidents, including systems that support operator actions, as well as the impact on containment integrity, should be taken into account."<br><br>This text gives a very important message; however, in practice, there are many errors observed when hazards are included in the Level-1/2 PSA model (E.g. many systems important for Level 2 PSA are not modelled in Level 1 PSA and thus internal or external events Level 1 PSA features are not reflected in the models for such systems developed within Level 2 PSA). | To elaborate in more detail recommendations for inclusion of hazards in the Level 1 and 2 PSA models. |

| | Issue | Proposed change |
|---|---|---|
| 9 | **Section 5 "Accident progression and containment analysis"**<br>• CET questions are presented in the form suitable for "EVENTTREE" code. This code is rarely used now as it does not allow the integrated model to be developed. No information on CET structure and nodal questions used in more common software (RiskSpectrum, Cafta, Saphire, FinPSA, NUPRA, etc.).<br>• No recommendations on how hazards, various radioactive sources, and operational states are represented in CET models.<br>• Very limited discussion on CET quantification (only para 5.26): There are many real problems in CET quantification, dealing with the size of the model, with the treatment of highly probable "success" branches, with the possibility to use 3 or more branches in CET logic, etc. | • To give clear recommendations on how end states in CET are defined.<br>• To provide recommendation for CET construction using integrated Level 1 and Level 2 PSA models.<br>• To provide examples of CETs developed using integrated models.<br>• To provide specific recommendations on specific features of CETs for PDSs developed for hazards and shutdown states (if any).<br>• To add a separate section "CET quantification" (as it is done in SSG-3 (Rev. 1)). |
| 10 | Paragraph 5.13. of SSG-4 states:<br><br>"The analyses of severe accidents should cover all sequences leading either to a successful stable state, where sufficient safety systems have operated correctly so that all the required safety functions necessary to cope with the plant damage state have been fulfilled, or to a containment failure state."<br><br>The statement is too strong. It is not possible to cover ALL sequences. | Make the statement broader (e.g. representative sequences from those leading …). |
| 11 | Paragraph 5.16. of SSG-4 states:<br><br>"The term 'containment event tree' is adopted in most Level 2 PSAs, while 'accident progression event tree', involving a greater level of modelling, is less frequently used. The term 'containment event tree' is used throughout this Safety Guide."<br><br>It is arguable that APET has a greater level of modelling. Otherwise, why use a term that is associated with a method that might have a lower level of details. | Remove speculative text "involving a greater level of modelling". |

| | Issue | Proposed change |
|---|---|---|
| 12 | Paragraph 5.19. of SSG-4 states:<br><br>"Regarding chronology, it is both useful and common practice to divide the containment event tree into phases sequential in time, with the transitions between phases representing important changes in the issues that govern accident progression, such as:"<br><br>This text seems to be based on old practices when severe accidents were analysed in several phases. It is not clear for newcomers and might not be needed for experienced experts. Nowadays, when integrated codes allow running accidents from IE up to releases or even doses it is not essential how phases are defined. | Add relaxation explaining the reasons for several phases and the possibility to assess lesser phases (up to one). |
| 13 | Paragraph 5.20. of SSG-4 states:<br><br>"Examples of a typical structure and nodal questions of a containment event tree for a typical pressurized water reactor with a large, dry containment are provided in Table 6."<br><br>It is arguable. Table 6 presents the typical structure of input to EVENTREE code, but not for CET. | Rewrite in the way that Table 6 presents nodal questions for specific software, but the content of these nodal questions can be used for any CET. |
| 14 | Paragraph 5.21. of SSG-4 states:<br><br>"Actions for severe accident management should be reflected in the Level 2 PSA. Typically, the human actions credited in PSA are included in plant procedures and severe accident management guidelines."<br><br>The original text is not complete. In Level 2 PSA it is possible to consider systems recovery (repair) if the time to release is long. In particular for modern designs, it might be longer than 72 hours, for SPF it is typically longer than 72 hrs. | To add a recommendation that systems recovery which is not credited in Level 1 PSA can be credited in Level 2 PSA (if feasibility is confirmed). |
| 15 | Paragraph 5.38. of SSG-4 states:<br><br>"Each value within the range of values that the uncertain parameter can take on is associated with a probability, thereby creating a probability density function or probability distribution."<br><br>Uncertainty dealing with the probability of phenomenological events in the CET (e.g. probability of hydrogen detonation) might not be treated as parametric and actually, it might not be represented by probabilistic distribution. The probability itself is the measure of uncertainty. | To add discussion on how the uncertainty of phenomenological events needs to be treated in CET. This issue has to be discussed as it is not clear for many practitioners (e.g. cases when error factor 10 was assigned for the event with probability 0.8 were observed). |

| | Issue | Proposed change |
|---|---|---|
| 16 | Paragraph 5.43. of SSG-4 states (without the note):<br><br>"Results and insights gained from the quantification of containment event trees should be summarized and discussed. Results are often tabulated in the form of a so-called containment performance matrix ('C matrix'), which is a concise way of comparing the relative likelihood of the various outcomes of the containment event trees. The C matrix identifies the conditional probabilities C (m, n) that a release category 'n' can be realized, given a plant damage state 'm'. Uncertainty analysis leads to alternative sets of values of the elements of the C matrix."<br><br>Para 5.43 is the first paragraph where the term 'release category' is used. It is not useful to present something which is not defined. The last sentence of para 5.43 is not clear (how uncertainty analyses can give alternative sets of values?). | To elaborate definition of end states of CETs, how they are defined, attributes for the definition of release categories approaches for grouping sequences in RC, etc.<br>It is proposed that paragraphs 6.3–6.10 (with a major update) be introduced in Section 5. |
| 17 | Para 5.47 of SSG-4 states (without the note):<br><br>"Generally, for each of the selected release categories, one representative accident sequence is selected for which a source term is estimated on the basis of results obtained from other PSAs, or using plant specific calculations employing an appropriate computer code for estimating source terms for severe accidents, as discussed in Section 6 and Annex II."<br><br>It is not clear which scenario is representative, what needs to be represented (what common features?), why other PSAs can be used and what is meant by the term 'source term'. | To rewrite the paragraph and add missing definitions. |
| 18 | **Section 6 "Source terms for severe accidents"**<br>The Section use 'release category', 'CET end state', 'source term' terms. There is an impression that sometimes it is used as synonyms but seems to be different.<br>For Level 2 PSA the current trend is to run analyses from IE up to releases, there is no need to calculate source terms for release categories (whatever it means).<br>Release categories already include radiological content. Further grouping in source terms require consideration of other features important for dose assessment. | Both Sections 5 and 6 require complete revision taking into account the current state of knowledge. |

| | Issue | Proposed change |
|---|---|---|
| 19 | **Annex III "Sample outline of documentation"**<br>Recommendations related to system models are needed for Level 2 PSA (spray system, containment isolation, systems, important for severe accidents progression and post-melt behaviour (core catcher, ex-vessel flooding, etc.).<br><br>Section M5 has confusing and unneeded details. Not clear what needs to be discussed in M5.1.3 "Summary of point estimate results for plant damage states analysed" section.<br><br>Inconsistencies of section M6 with Section 5 and 6 of SSG-4. The SFP, shutdown, hazards PSA are not reflected in the structure (except for Section 3). | To comprehensively review and update the proposed outline. The SFP, shutdown state, hazards PSA need to be reflected in the outline. |
| 20 | Important information can be presented in additional Annexes. | To add recommendations on approaches for assessment of the reliability of passive systems in Level 2 PSA (core catcher, recombines, etc.).<br>Include recommendations or examples of approaches for assessment of certain important phenomena (grip rupture, DCH, detonation/deflagration, etc.). Include illustrations for bridge trees and CET.<br>In the new Annex add development and use of Level 2 PSA at the design stage. |

## LIST OF PARTICIPANTS

| | |
|---|---|
| Abdelaal, M. | Egyptian Atomic Energy Authority, Egypt |
| Abe, Y. | Nuclear Regulation Authority, Japan |
| Ahangari Shahdehi, R. | Atomic Energy Organization of Iran-Nuclear Science and Technology Research Institute, Islamic Republic of Iran |
| Ahmed, S. | Bangladesh Atomic Energy Commission, Bangladesh |
| Ahmed Abd El-Menem El-Said El-Baz, M. | Nuclear Power Plants Authority, Egypt |
| Alsuwaidi, J. | Federal Authority for Nuclear Regulation, United Arab Emirates |
| Astakhov, V. | OKB Gidropress JSC, Russian Federation |
| Basov, D. | OKB Gidropress JSC, Russian Federation |
| Bayle, J. | Électricité de France, France |
| Biro, M. | United States Nuclear Regulatory Commission, United States of America |
| Bukhari, S.M.W. | Pakistan Nuclear Regulatory Authority, Pakistan |
| Campos, J.M. | Autoridad Regulatoria Nuclear, Argentina |
| Chirila, M. | Canadian Nuclear Safety Commission, Canada |
| Cho, J. | Korea Atomic Energy Research Institute, Republic of Korea |
| Coelho Maturana, M. | Amazônia Azul Tecnologias de Defesa S.A., Brazil |
| Coqueijo De Abreu, C. | Brazilian Navy, Brazil |
| Correa De Guama, D. | Eletrobras Eletronuclear, Brazil |
| Delcausse-Malbec, F. | Électricité de France, France |
| Delvax, J. | BEL V, Belgium |
| Elbordany, A. | Egyptian Nuclear & Radiological Regulatory Authority, Egypt |
| Garcia, J.M. | National Atomic Energy Commission, Argentina |
| Gasparov, D. | OKB Gidropress JSC, Russian Federation |
| Gimenez, M. | National Atomic Energy Commission, Argentina |
| Golakoti, S. | Office for Nuclear Regulation, United Kingdom of Great Britain and Northern Ireland |

| | |
|---|---|
| Gonzalez Rodriguez, V. | Comisión Nacional de Seguridad Nuclear y Salvaguardias, Mexico |
| Grindon, E. | Jacobs, United Kingdom of Great Britain and Northern Ireland |
| Harbachova, N. | State Scientific Institution "The Joint Institute for Power and Nuclear Research - Sosny", Belarus |
| Harman, N. | Jacobs, United Kingdom of Great Britain and Northern Ireland |
| Holmberg, J.E. | Radiation and Nuclear Safety Authority, Finland |
| Hrytskevich, D. | Republican Unitary Enterprise "Belarusian Nuclear Power Plant", Belarus |
| Hyros, J. | Slovenske elektrarne a.s. Nuclear Power Plant Bohunice, Slovakia |
| Ibarra, V.A. | Autoridad Regulatoria Nuclear, Argentina |
| Islam, M. | Bangladesh Atomic Energy Commission, Bangladesh |
| Ivochkin, M. | Scientific and Engineering Centre for Nuclear and Radiation Safety, Russian Federation |
| Jafarian, R. | Iran Nuclear Regulatory Authority, Islamic Republic of Iran |
| Julin, A. | Radiation and Nuclear Safety Authority, Finland |
| Kanetsyan, G. | Nuclear and Radiation Safety Centre, Armenia |
| Karsa, Z. | NUBIKI Nuclear Safety Research Institute Ltd., Hungary |
| Kiliç, F. | Nuclear Regulatory Authority, Türkiye |
| Kioseyian-Pontillo, G. | Institut de Radioprotection et de Sûreté Nucléaire, France |
| Kojo, R. | Nuclear Regulation Authority, Japan |
| Kovacs, Z. | RELKO Ltd., Slovakia |
| Kubicek, J. | UJV Rez, a.s., Czech Republic |
| Kubo, S. | Japan Atomic Energy Agency, Japan |
| Kudrytskyi, O. | State Enterprise "State Scientific and Technical Center for Nuclear and Radiation Safety", Ukraine |
| Kumagai, Y. | OECD, Nuclear Energy Agency |
| Laato, T. | Fennovoima Oy, Finland |
| LaBarge, N. | Westinghouse Electric Company, LLC, United States of America |

| | |
|---|---|
| Lajtha, G. | NUBIKI Nuclear Safety Research Institute Ltd., Hungary |
| Levin, V. | Joint Stock Company Atomenergoproekt, Russian Federation |
| Linthicum, R. | Pressurized Water Reactor Owners Group, United States of America |
| Lityshev, A. | OKB Gidropress JSC, Russian Federation |
| Liubarskii, A. | Joint Stock Company Atomenergoproekt, Russian Federation |
| Lopez Garcia, A. | Comission Nacional de Seguridad Nuclear y Salvaguardias, Mexico |
| Luis Hernandez, J. | International Atomic Energy Agency |
| Mahmoud, A. | Nuclear Power Plants Authority, Egypt |
| Mancheva, K. | GCR Ltd., Bulgaria |
| Mikhalycheva, E. | State Scientific Institution "The Joint Institute for Power and Nuclear Research – Sosny", Belarus |
| Minibaev, R. | International Atomic Energy Agency |
| Mladenova, S. | Kozloduy NPP Plc., Bulgaria |
| Moraru, D. E. | Societatea National "Nuclearelectrica" SA Cernavoda NPP, Romania |
| Movsisyan, M. | Armenian Scientific-Research Institute for NPP Operation "ARMATOM" CJSC, Armenia |
| Mullin, E. | NuScale Power, United States of America |
| Nakamura, K. | Nuclear Risk Research Center, Central Research Institute of Electric Power Industry, Japan |
| Nikolaeva, A. | OKB Gidropress JSC, Russian Federation |
| Nishimura, T. | Nuclear Regulation Authority, Japan |
| Nunes Araújo, N. | Amazônia Azul Tecnologias de Defesa S.A., Brazil |
| Nusbaumer, O. | Leibstadt NPP Kernkraftwerk Leibstadt AG, Switzerland |
| Olofsson, J. | Ringhals, AB, Sweden |
| Patel, B. | Radiation Protection and Nuclear Safety, European Commission |
| Petkov, G. | Kozloduy NPP Plc., Bulgaria |
| Poghosyan, S. | International Atomic Energy Agency |

| | |
|---|---|
| Powell, M. | Arizona Public Service Company, Pressurized Water Reactor Owners Group, United States of America |
| Rahni, N. | Institut de Radioprotection et de sûreté nucléaire, France |
| Raimond, E. | Institut de Radioprotection et de sûreté nucléaire, France |
| Ranguelova, V. | International Atomic Energy Agency |
| Reka, A. | JSC Atomproekt, Russian Federation |
| Rodriguez Gual, M. | Amazônia Azul Tecnologias de Defesa S.A., Brazil |
| Romberg Junior, H C. | Amazônia Azul Tecnologias de Defesa S.A., Brazil |
| Rzentkowski, G. | International Atomic Energy Agency |
| Sargsyan, S. | Haykakan Atomayin Electrakayan CJSC, Armenia |
| Schneider, R. | Westinghouse Electric Corporation, United States of America |
| Sedlazek, L. | International Atomic Energy Agency |
| Segovia, W.F. | National Atomic Energy Commission, Argentina |
| Siklosi, A. | Hungarian Atomic Energy Authority, Hungary |
| Siklossy, T. | NUBIKI Nuclear Safety Research Institute Ltd., Hungary |
| Spenlinger, R. | RELKO Ltd., Slovakia |
| Stebenev, N. | JSC Atomproekt, Russian Federation |
| Summit, R. | Engineering Planning and Management, United States of America |
| Tyrväinen, T. | VTT Technical Research Centre of Finland, Finland |
| Ui, A. | Nuclear Risk Research Center, Japan |
| Vesecky, R. | Elektrarna Dukovany II a.s., Czech Republic |
| Williams, C. | NuScale Power, United States of America |
| Yalaoui, S. | Canadian Nuclear Safety Commission, Canada |
| Yamano, H. | Japan Atomic Energy Agency, Japan |
| Yang, Y. | Shanghai Nuclear Engineering Research and Design Institute Co. Ltd., China |
| Zaidi, I. | Pakistan Atomic Energy Commission, Pakistan |
| Zakusilov, V. | JSC Atomproekt, Russian Federation |

## CONTRIBUTORS TO DRAFTING AND REVIEW

| | |
|---|---|
| Bayle, J. | Électricité de France, France |
| Bukhari, S.M.W. | Pakistan Nuclear Regulatory Authority, Pakistan |
| Campos, J.M. | Autoridad Regulatoria Nuclear, Argentina |
| Cho, J. | Korea Atomic Energy Research Institute, Republic of Korea |
| Coelho Maturana, M. | Amazônia Azul Tecnologias de Defesa S.A., Brazil |
| Delcausse-Malbec, F. | Électricité de France, France |
| Garcia, J.M. | National Atomic Energy Commission, Argentina |
| Gimenez, M. | National Atomic Energy Commission, Argentina |
| Holmberg, J.E. | Radiation and Nuclear Safety Authority, Finland |
| Ibarra, V.A. | Autoridad Regulatoria Nuclear, Argentina |
| Ivochkin, M. | Scientific and Engineering Centre for Nuclear and Radiation Safety, Russian Federation |
| Julin, A. | Radiation and Nuclear Safety Authority, Finland |
| Kanetsyan, G. | Nuclear and Radiation Safety Centre, Armenia |
| Kioseyian-Pontillo, G. | Institut de Radioprotection et de Sûreté Nucléaire, France |
| Kovacs, Z. | RELKO Ltd., Slovakia |
| Kubo, S. | Japan Atomic Energy Agency, Japan |
| LaBarge, N. | Westinghouse Electric Company, LLC, United States of America |
| Linthicum, R. | Pressurized Water Reactor Owners Group, United States of America |
| Lityshev, A. | OKB Gidropress JSC, Russian Federation |
| Liubarskii, A. | Joint Stock Company Atomenergoproekt, Russian Federation |
| Luis Hernandez, J. | International Atomic Energy Agency |
| Minibaev, R. | International Atomic Energy Agency |
| Moraru, D. E. | Societatea National “Nuclearelectrica” SA Cernavoda NPP, Romania |
| Nakamura, K. | Nuclear Risk Research Center, Central Research Institute of Electric Power Industry, Japan |

| | |
|---|---|
| Nunes Araújo, N. | Amazônia Azul Tecnologias de Defesa S.A., Brazil |
| Nusbaumer, O. | Leibstadt NPP Kernkraftwerk Leibstadt AG, Switzerland |
| Petkov, G. | Kozloduy NPP Plc., Bulgaria |
| Rahni, N. | Institut de Radioprotection et de sûreté nucléaire, France |
| Raimond, E. | Institut de Radioprotection et de sûreté nucléaire, France |
| Reka, A. | JSC Atomproekt, Russian Federation |
| Rodriguez Gual, M. | Amazônia Azul Tecnologias de Defesa S.A., Brazil |
| Romberg Junior, H C. | Amazônia Azul Tecnologias de Defesa S.A., Brazil |
| Schneider, R. | Westinghouse Electric Corporation, United States of America |
| Segovia, W.F. | National Atomic Energy Commission, Argentina |
| Spenlinger, R. | RELKO Ltd., Slovakia |
| Stebenev, N. | JSC Atomproekt, Russian Federation |
| Stone, Z. | International Atomic Energy Agency |
| Summit, R. | Engineering Planning and Management, United States of America |
| Tyrväinen, T. | VTT Technical Research Centre of Finland, Finland |
| Ui, A. | Nuclear Risk Research Center, Japan |
| Yalaoui, S. | Canadian Nuclear Safety Commission, Canada |
| Yamano, H. | Japan Atomic Energy Agency, Japan |
| Zakusilov, V. | JSC Atomproekt, Russian Federation |

IAEA
International Atomic Energy Agency

# CONTACT IAEA PUBLISHING

Feedback on IAEA publications may be given via the on-line form available at:
www.iaea.org/publications/feedback

This form may also be used to report safety issues or environmental queries concerning IAEA publications.

Alternatively, contact IAEA Publishing:

Publishing Section
International Atomic Energy Agency
Vienna International Centre, PO Box 100, 1400 Vienna, Austria
Telephone: +43 1 2600 22529 or 22530
Email: sales.publications@iaea.org
www.iaea.org/publications

Priced and unpriced IAEA publications may be ordered directly from the IAEA.

## ORDERING LOCALLY

Priced IAEA publications may be purchased from regional distributors and from major local booksellers.

25-00454E

FSC
www.fsc.org
MIX
Papier | Fördert
gute Waldnutzung
FSC® C083411

Zeitfracht Medien GmbH
Ferdinand-Jühlke-Straße 7
99095 Erfurt, Deutschland
produktsicherheit@kolibri360.de